Interdisciplinary Lively Application Projects

Modules

Tools for Teaching 2000

published by

The Consortium for Mathematics
and Its Applications, Inc.
Suite 210
57 Bedford Street
Lexington, MA 02420

edited by

Paul J. Campbell
Campus Box 194
Beloit College
700 College Street
Beloit, WI 53511–5595
campbell@beloit.edu

289350 AUG 1 9 2004

ISBN 0–912843–72–1
Typeset and printed in the U.S.A.

Table of Contents

Introduction: More Tools for the Toolbox

Did you ever reach into your toolbox and find something surprising or new that actually looked like the right tool for the job at hand? That is a pleasant situation that becomes even better when the tool actually does the job. We hope that happens to you as you reach into this issue of *Tools for Teaching*. Whether small-group projects are new tools or old tools for your courses, you probably didn't expect to find them in this toolbox, but they are here. We hope that you find them helpful in accomplishing your course goals. We hope that they help your students learn and grow. We hope that you become excited and build new tools for others or share your old ones.

The new elements (tools) in this year's *Tools for Teaching* issue are called *Interdisciplinary Lively Application Projects (ILAPs)*. Three articles in this issue provide information about these teaching tools and the roles that they play in the teaching and learning of mathematics.

The articles also describe the activities of Project INTERMATH, which is a National Science Foundation–supported consortium working with COMAP to change the academic culture of undergraduate mathematics, science, and engineering education. Specifically, the NSF initiative "Mathematical Sciences and their Applications Throughout the Curriculum" seeks to improve the educational culture through the enhancement of interdisciplinary cooperation and coordination. The primary activities of Project INTERMATH are development of

- ILAPs,

- integrated and/or interdisciplinary curricula as national models, and

- faculty to engage in interdisciplinary activities and cultural change.

The articles in this volume address these activities.

ILAPs are small group projects that are developed through interdisciplinary cooperation of faculty from partner disciplines. These projects weld mathematics with the concepts and principles of the other disciplines and enable faculty to accomplish a variety of course goals. From the student's perspective, ILAPs motivate the need to develop mathematical concepts and skills, provide interest in further study of mathematics and other disciplines, and produce a broader outlook on problem solving.

ILAPs are problem-solving projects and therefore are different from UMAP Modules. UMAP Modules have been the usual content of past *Tools for Teaching* volumes, and they will appear with ILAPs in future issues of *The UMAP Journal* and *Tools for Teaching* volumes. UMAP Modules are teaching/learning materials that contain explanations of content ideas and concepts, guided learning experiences, examples and exercises, and often an exam on the material. ILAPs, on the other hand, contain few if any explanations, examples, or exams.

This volume contains classroom resource material (tools for teaching) in the form of ILAPs. An ILAP contains

- background information and introduction to the situation,

- project handouts,

- usually specific requirements to be fulfilled in a solution, and

- other supporting materials (sample solutions, further background material, notes for the instructor) as appropriate.

These projects are formatted much like case studies. They often require students to use scientific and quantitative reasoning, mathematical modeling, symbolic manipulation, and computation to solve problems, analyze scenarios, understand issues, and answer questions.

The student effort can involve individual or group work, and the final product is an oral presentation or a written report. The problems are intended to be relevant to students' interests and often involve open-ended components.

The level of prerequisite skills varies from algebra and precalculus to higher-level undergraduate concepts in calculus, differential equations, or probability and statistics.

To be effective course materials, ILAPs must be of high quality in both content and presentation. COMAP invites you to become involved with the development and use of ILAPs:

- Use these projects in your classes.

- Modify them as needed to mesh with your course and your students.

- Write your own ILAPs with partner faculty from other disciplines and submit them to *The UMAP Journal*.

Follow the normal directions in the Guide for Authors section of this issue to submit an ILAP for publication or else send three copies directly to the ILAP editor:

> Chris Arney
> Dean of the School of Mathematics and Sciences
> The College of Saint Rose
> 432 Western Avenue
> Albany, NY 12203

A key requirement is that an ILAP must have at least two author-collaborators, preferably from different disciplines. The project itself must provide a broad interdisciplinary experience for the students engaged in its solution.

We hope that

- the authoring, editing, and use of ILAPs will take place in a productive setting of cooperation and partnership; and

- ILAPs will enjoy similar success in terms of production, longevity, availability, and use as the UMAP Modules, which were first published in 1977 and are still being written and used today.

Good luck in reaching into your *Tools for Teaching* toolbox! May you find something surprising or new that actually does the job that you need done in teaching your students.

— Chris Arney

Acknowledgment

The production of this volume would not have been possible without the organizational and computer expertise of MAJ Kirk Benson of the U.S. Military Academy, now at Georgia Institute of Technology.

Overview of Project INTERMATH: Seeking Cultural Change

Project INTERMATH Staff

Goal

The goal of Project INTERMATH is to inspire and lead mathematics and its partner disciplines in undergraduate education to improve the academic culture through more interdisciplinary learning and cooperation. The processes used by the INTERMATH consortium to achieve this goal are the

- development and use of Interdisciplinary Lively Application Projects (ILAPs),

- design and implementation of integrated and/or interdisciplinary curricula in mathematics programs, and

- development of faculty to collaborate with colleagues in other disciplines to provide broad, relevant learning experiences for students.

In plain words, we seek to change how we teach and communicate to improve student learning. We believe the processes of developing and using ILAPs is the engine that drives the success of this project. These processes generate communication, involvement, and connection among disciplines, which features are needed for educational improvement and cultural change.

ILAPs

Students benefit from working in teams on ILAPs because they work in concert to formulate and solve relevant and interesting problems rather than compete against each other to achieve a disciplinary standard. Communication and confidence are improved through group work. In the group and team setting, students feel more open to trying new approaches and being creative in their problem solving.

Mathematical modeling and scientific problem-solving are two of the most challenging tasks for students to learn, and group work provides an excellent forum for those endeavors. ILAPs often encourage the use of computer or calculator technology and hence develop those important skills. ILAPs can reveal connections among courses, disciplines, and ideas.

Most important, these projects motivate and inspire student engagement in their learning and are opportunities for students to construct their own mathematical capacity through discovery learning and creative thinking. Since ILAPs necessitate that students take responsibility for their learning, they can be valuable in the development of student-centered courses.

ILAPs are more than just products; they are the means by which faculty members and students from different disciplines become scholarly partners. The development or use of ILAPs is an empowering experience for faculty members. They require communication and collaboration between faculty from several disciplines. They consider pedagogical issues as they design the student experience for the project activities. The ownership of the elements of the project often brings additional enthusiasm to the classroom. Once faculty members experience the benefits of ILAPs, they use them more and shift emphasis from topic coverage to student-centered learning. The faculty improve the way they teach, they involve themselves in interdisciplinary collaboration, and they ultimately change the culture at their school. The artificial walls between disciplines, which can act as barriers for student learning, begin to disappear, and new paths to academic engagement appear.

Curricular Initiatives

Project INTERMATH has success in the design and implementation of integrated and/or interdisciplinary curricula in mathematics programs. In several cases, national models have been proposed to address issues of ABET (Accreditation Board for Engineering and Technology) accreditation, college algebra reform, and the effective inclusion of scientific computing. At some schools, ILAPs have ignited the process of collaboration on curricular issues. In other cases, curricular changes have proceeded the development of ILAPs.

As faculty partners co-design and coordinate curricula, new opportunities become available for students, involved faculty members develop as scholars and educators, and the culture of the programs transforms. Some of the student growth opportunities realized in this project are:

- seeing the relevance of mathematics,

- appreciating the multifaceted nature of problem solving,

- learning the role of mathematics as a structure for scientific reasoning,

- experiencing verbal and written communication with different media, and

- working cooperatively in groups.

Faculty development can take place in areas like

- teamwork,

- knowledge and appreciation of other disciplines,

- linking of disciplines,

- innovative pedagogy,

- curricular needs of programs,

- empowerment as teachers, and

- student growth needs.

Finally, cultural changes can take the form of

- new shared ownership of programs and student development,

- better interdisciplinary and interdepartmental communication and coordination,

- new innovation in programs and pedagogy, and

- better understanding of student and program needs.

Faculty Development

Project INTERMATH's faculty development program of workshops and conferences has served several purposes:

- spreading information about ILAPs and curricular models,

- introducing faculty to interdisciplinary collaborations,

- developing and using ILAPs,

- adapting and implementing new curricular ideas, and

- confronting the barriers to interdisciplinary cooperation.

As in any learning situation, follow-on activity to the workshop is necessary to ensure that participants act on the information they received and the plans they made at the workshop. This is particularly necessary when the ultimate goal is to change a culture by breaking old habits and instilling new methods. This follow-on activity is tailored to specific institutions, since the existing cultures and barriers differ from institution to institution. Producing academic cultural change is slow work, requires a long-term commitment, and is essentially an individual conversion process.

Website

Project INTERMATH has a website at http://www.projectintermath.org that houses:

- ILAPs and ideas for using the materials in courses,

- links to other ILAP materials,

- INTERMATH curriculum models,

- references to related NSF funding opportunities for course development,

- Project INTERMATH newsletters,

- links for modeling and interdisciplinary teaching,

- INTERMATH workshop schedule and links to NSF workshop schedule, and

- links to other interdisciplinary projects.

Interdisciplinary Contest in Modeling (ICM)

Building upon the successful Mathematical Contest in Modeling (MCM), Project INTERMATH in collaboration with COMAP initiated an interdisciplinary contest for undergraduate teams in February of 1999.

Teams of three students work over a weekend on an interdisciplinary problem and submit a written report on their solution and findings. The discipline for the first three contests and for next year's contest is *environmental science*. The first contest consisted of a pollution problem involving detection and monitoring from chemical data from a real set of wells. The second contest was managing elephant herds on game preserves in Africa. This past year, the problem examined the infestation of an exotic water pest, the zebra mussel.

The "Outstanding" papers, along with expert commentary, are published annually in an issue of *The UMAP Journal*. Look for this year's results in a special issue devoted to the ICM. The directors of the contest hope that this contest exposes students to interdisciplinary problem-solving and promotes the goals of Project INTERMATH and the NSF initiative.

Summary

In many schools and programs involved with ILAPs and interdisciplinary curricula as part of Project INTERMATH, students' skills in technology, communication, modeling, and reasoning are improving. Faculty members are enthused and involved as interdisciplinary partners, sharing ideas and helping each other. Cultural change is taking place, and a more interdisciplinary perspective is emerging. We find the "center of gravity" changing.

Integrated and Interdisciplinary Curricula

Project INTERMATH Staff

Introduction

The undergraduate mathematics curriculum has often been identified as the "gate-keeper" to America's future in science and technology—a "filter" that too often deflects students into nonquantitative studies. Academia should be a "caretaker" and "developer," not a "gatekeeper." The National Science Foundation (NSF) has recently led reforms to improve the teaching and content of mathematics at the undergraduate level. Much of this reform occurred in calculus.

However, to realize the objective of transforming the "filter" into a "pump" requires changes in the culture of the mathematics, science, and engineering teaching communities. A goal of Project INTERMATH, through its funding by NSF, is curricular and cultural reform through the development and use of student projects that integrate the curricula of mathematics and partner disciplines—a cultural change. This paper explains some of the ways that Project INTERMATH is achieving this goal.

It can be done. At many schools involved in INTERMATH activities, communication among disciplines has improved dramatically, as partner disciplines provide input to the design of the mathematics curriculum as well as participate in its presentation. The development of growth models and outcome goals in the INTERMATH integrated/interdisciplinary curricula help students to visit essential concepts and applications at increasingly more sophisticated levels as the curriculum unfolds. Concurrent and downstream partner disciplines build upon student outcomes by revisiting and extending the student growth into their courses and programs.

How can it be done? INTERMATH promotes curricular and cultural reform through an interdisciplinary approach to student development that requires collaborative and innovative curriculum design. Faculty members are empowered to develop and use relevant student projects that link mathematics with partner disciplines. These projects are called Interdisciplinary Lively Application Projects (ILAPs). The mathematics classroom and the homework venue become an interactive laboratory of interdisciplinary discovery. Students witness the relevance of mathematics by solving current problems designed and presented by practitioners from partner disciplines. Many of these projects are revisited in downstream higher-level courses. Much like the role of examples in the following quote, ILAPs become an important ingredient in student learning.

> A good stock of examples, as large as possible, is indispensable for a thorough understanding of any concept, and when I want to learn something new, I make it my first job to build one.
>
> —Paul R. Halmos

Partners in the process. Several schools in the INTERMATH consortium have developed integrated or interdisciplinary curricula. Others have developed ILAPs and conducted various interdisciplinary activities. The INTERMATH consortium partners with COMAP for administrative, publishing, and curricular support. The schools in the consortium are:

Carroll College (Montana)	Oklahoma State University
Clark-Atlanta University	Prairie View A&M University
Francis Marion College	Texas Southern University
Georgia College and State University	Tulane University
Harvey Mudd College	United States Military Academy
Huston-Tillotson College	University of Redlands
Macalester College	Wiley College

Other schools involved in INTERMATH activities include:

Benedict College	a consortium of Tribal Colleges in Montana
College of Saint Rose	Trinity College
Lafayette College	University of Colorado at Denver
Philander-Smith College	University of Maryland
Smith College	University of Tulsa

We summarize below examples of new curricular development projects of five consortium schools (United Stated Military Academy, Carroll College, Harvey Mudd College, Macalester College, and Texas Southern University) to provide a flavor of this type of curricula.

ABET 2000 Standards

The two-year core curricular programs developed at three schools (United Stated Military Academy, Carroll, Harvey Mudd) and presented below are particularly appropriate for schools seeking accreditation by the Accreditation

Board for Engineering and Technology (ABET). Most undergraduate engineering programs have only 15–18 semester hours available for their core mathematics sequence, which traditionally has consisted of three semesters of calculus, a semester of differential equations, and an "engineering mathematics" course, possibly tailored to the specific engineering discipline.

ABET 2000 contains outcomes-based standards that simply specify that the graduates of the program possess

- an ability to apply knowledge of mathematics, science, and engineering; and

- an ability to design and conduct experiments, as well as to analyze and interpret data.

Some of the disciplinary societies amplify these basic standards. For example, civil engineering specifies that

- Graduates must demonstrate a proficiency in mathematics through differential equations, probability and statistics, calculus-based physics and general chemistry.

We believe that the three INTERMATH core curricula presented below address the needs of ABET programs as they treat a broad set of integrated topics (including the addition of probability and statistics, matrix algebra, and discrete mathematics) and through their interdisciplinary approach connect well with the science courses in the programs. Equally important for other ABET outcomes and standards, these curricula also address student growth in the dimensions of communication, computation, teamwork, independent learning, and interdisciplinary perspectives.

United States Military Academy

West Point requires all of its students to complete a four-course mathematical sciences core that integrates the content of calculus I, calculus II, calculus III, discrete mathematics, linear algebra, differential equations, and probability and statistics. USMA has successfully taught this "7-into-4" curriculum to thousands of cadets, with positive reactions from both the faculty in the mathematical sciences and the partner disciplines of engineering, physical sciences, and economics.

The following four courses comprise the core mathematics curriculum for all students during the first four semesters at the Academy:

- **Discrete Dynamical Modeling and an Introduction to Calculus.** Difference equations through eigenvalues for systems and a transition from sequences to continuous functions in calculus. This course, containing considerable modeling, is linked to the concurrent chemistry and computer science courses through interdisciplinary projects.

- **Calculus I and Differential Equations.** Differential and integral calculus with considerable study of differential equations and applications involving change, including systems of differential equations solved with matrix algebra techniques.

- **Calculus II.** Multivariable calculus with vectors. This course has strong links to the concurrent physics and economics courses through interdisciplinary projects (see the Mathematics/Physics section below).

- **Probability and Statistics.** Discrete and continuous probability distributions, confidence intervals, and hypothesis testing. A capstone project in this course assesses students' progress in the math-science-and-technology components of the USMA program.

While the four courses listed indicate the major topics of the semester, the curriculum regularly compares and contrasts the relationship among the following mathematical structures:

- discrete and continuous,

- linear and nonlinear,

- deterministic and stochastic.

Interdisciplinary Projects

An important activity of Project INTERMATH at West Point is the development and use of ILAPs. These small group projects are developed through the cooperation of faculty from mathematics and partner disciplines. These projects have the goals of welding mathematics with the concepts and principles of partner disciplines, promoting curricular advancement by focusing on student growth, and promoting faculty growth through the ILAP development processes.

From the student's perspective, ILAPs provide applications that motivate mathematical concepts and skills; provide interest in other subjects that become accessible through further study of mathematics; and enable a broader, more interdisciplinary outlook.

ILAPs are interdisciplinary in their nature, terminology, and notation. They usually require students to use scientific and quantitative reasoning, mathematical modeling, symbolic manipulation, and computation to explore, solve, and analyze scenarios, issues, and questions. The student's work is presented in oral or written reports. The actual time spent by a student to solve and present an ILAP can be difficult to predict, but most of the USMA ILAPs require 6–8 hours of student effort.

The faculty at West Point works together in order to enable cadets to develop confidence and competence in using mathematics in a variety of settings. Also, the impact of technology on teaching and learning has helped to increase the student learning opportunities. Interdisciplinary activities have become a standard method of presenting learning opportunities for the Academy's students.

Mathematics / Physics

West Point has developed a semester-long interdisciplinary course that combines multivariable calculus and physics for second-year students. In an effort to enhance the students' understanding of material, faculty members from Mathematics and from Physics decided that it would be effective to team up and present an interdisciplinary math/physics class. One of the goals for this new class is to develop mathematical relationships in a richer context than has been presented in the past.

Carroll College of Montana

Recognizing the dramatic impact that technology is having on applied mathematics, the Carroll College Dept. of Mathematics, Engineering, and Computer Science has been involved in formulating an innovative curriculum that is designed to engage students in the art of mathematics while focusing on a wide array of applications.

This effort has resulted in changes both in what is taught in the mathematics classroom and how it is taught. Through involvement in Project INTERMATH, Carroll has developed its new curriculum in consultation with others in the consortium, and this association has proved very fruitful. In particular, West Point's 7-into-4 curriculum and ILAPs, coupled with COMAP's innovative text, *Principles and Practice of Mathematics* [COMAP 1996], have provided models and ideas upon which to build the curriculum. The following items describe the key features of the program.

Activities

Carroll's curriculum project consists of four parts:

- To refine and improve the integrated mathematics curriculum for freshmen and sophomores and to extend the integrated program into the junior and senior years.

- To meet better the needs of students in other disciplines by tailoring service courses.

- To develop stronger connections with other disciplines (engineering, science, computer science, and business/economics) by jointly creating ILAPs and promoting the shared use of ILAPs.

- To establish connections among all disciplines by promoting the cooperative development and use of ILAPs.

Courses

Carroll College's core program is an integrated sequence during the first two years consisting of four 5-semester-hour courses. The sequence is required for all mathematics and engineering majors. Topics include:

- **Course 1:** Difference Equations (1 credit), Calculus (3 credits), and Differential Equations (1 credit);

- **Course 2:** Linear Algebra (2 credits) and Calculus (3 credits);

- **Course 3:** Multivariable Calculus (4 credits) and Differential Equations and Differential Equations (1 credit); and

- **Course 4:** Linear Algebra (1 credit), Systems of Difference and Differential Equations (2 credits), and Probability and Statistics (2 credits).

Integrating the computer science course, providing connections with partner disciplines, and the cooperative design and mutual integration of ILAPs bring from diverse fields an interdisciplinary flavor to the four courses. Carroll continues to institutionalize its cultural change across the disciplines and was recently certified by ABET under the new ABET 2000 criteria.

Integrated

The goal is to present mathematics as a unified topic, so as to preserve the beauty and integrity of the subject. By focusing on the solution of meaningful applications, this integration evolved naturally.

For example, the first-year calculus course begins with modeling problems using difference equations. A study of difference equations and systems of difference equations leads naturally to concepts in calculus, differential equations, and linear algebra. Linear algebra can be used to integrate topics in abstract algebra and transformational geometry. Concepts in probability and statistics arise naturally in the study of moments and in error analysis.

Because of use of technology, fundamentals of computer programming are presented in the context of problem solving. In general, mathematical topics are integrated throughout the courses for both the major and the non-major.

Interdisciplinary

Particularly through the use of ILAPs, the study of mathematics is driven by applications from a wide variety of disciplines, including business and finance, science, engineering, environmental studies, medicine, social sciences, and computer science. These often provide the basis for further explorations in history, ethics, political science, art, literature, and theology, among others. Emphasis on combining mathematics with other disciplines is made evident by requiring students majoring in mathematics to complete a concentrated program of study in another discipline such as engineering, secondary education, business, biology/chemistry, or computer science.

Technology Intensive

Powerful calculators and computer technology have given Carroll's students the power to extend understanding of topics and to solve advanced problems. Symbolic-graphing calculators, computer laboratories, information systems, computer algebra systems, statistics packages, spreadsheets, and calculator-driven instrumentation are used extensively. Classes meet regularly in up-to-date computer labs using varied software that provides problem-solving, learning, and visualization tools. With calculator-based instrumentation, real data are collected for mathematical modeling projects. The campus network is connected to the Internet for access to curriculum resource materials, library resources, and databases.

Collaborative and Professionally Oriented

Students work collaboratively in groups on in-class discovery and exploration exercises, in the computer labs on laboratory assignments, and outside of class to prepare written reports and oral presentations on applications projects. Formal written reports and oral presentations are a part of each student project. Presentations are made to classmates and the teacher and to other interested students and faculty, and students take ownership of their work and are proud to present it to others. They develop skills and attitudes that will serve them well in professional careers. For example, Carroll's students in mathematics secondary education completed a Web-based project and a thesis on "Mathematics and Music" and presented their results at regional and national meetings.

A New Spirit in Students

Students now show enthusiasm for learning and are proud to be in the mathematics program. They enjoy the opportunities to exercise their creativity. Their spirit is contagious; Carroll has more students majoring in mathematics now than before. From the faculty perspective, teaching is even more rewarding and enjoyable. Carroll now has cooperative efforts with local school districts to team-teach its first-year mathematics course. This gives students a rigorous and rewarding experience in mathematics while earning college credits in high school.

Harvey Mudd College

Harvey Mudd's Mathematics Dept. has put into place a new approach for teaching the engineering mathematics curriculum (hereafter, Core Math). The previous mathematics sequence was traditional, with courses in calculus, linear algebra, and ordinary differential equations (ODEs).

A distinctive feature of Harvey Mudd College (HMC) is that a year of high school calculus is required for matriculation. HMC's project relies on this accelerated background of the students; this approach could be implemented at other institutions in sections that had high school calculus as a prerequisite.

The core curriculum requires more open-ended, project-oriented work and makes a natural bridge to the ILAP development. HMC requires that students have at least two modeling projects each semester, with a written technical report. Students prepare short oral presentations of their modeling projects and prepare a poster. At the end of each semester, students conduct a "poster session" wherein they display the results of their projects to faculty and other students.

Objectives

There are several objectives that define this approach to revising Core Math:

- improve coordination with science and engineering courses,

- increase the opportunities for reinforcement of concepts,

- reduce compartmentalization of mathematical ideas,

- introduce elementary probability and statistics,

- introduce more creative and modern mathematical applications,

- use technology to enhance the learning opportunities for students, and

- provide more varied student experiences, such as project oriented group work and presentations.

To accomplish these objectives, HMC refashioned its four-semester sequence of Univariable Calculus, Multivariable Calculus, Linear Algebra, and Differential Equations into eight half-courses, or Modules. These Modules present the material in a nontraditional order and with novel emphases. Coursework in the Modules is enhanced by Web-based tutorials and quizzes.

Modules

There are eight Modules, each about seven weeks long. Modules are taken as half-courses and are graded and recorded independently on a student's transcript.

- Module 1 (Freshman Fall): **Univariable Calculus.** As described above, the premise is that students will have had one year of calculus in high school. Module 1 is a review of basic concepts and techniques of calculus, with the following difference: complex numbers and complex functions are used strategically to enhance the course and make it seem different from their high school experience. For example, sequences and series are presented using complex numbers, and power series of a complex variable are introduced. Anticipating the line integral of multivariable calculus, simple contour integrals are studied.

- Module 2 (Freshman Fall): **Linear Algebra and Discrete Dynamics.** An introduction to matrices and matrix algebra, using discrete dynamical systems to motivate the mathematics and for applications. Systems of equations, matrix transformations, linear independence, determinants, introduction to eigenvalues and eigenvectors.

- Module 3 (Freshman Spring): **Introduction to Ordinary Differential Equations.** First- and second-order differential equations, with applications. Emphasis is on solution methods and applications to scientific and engineering problems. Numerical methods and using computational tools for exploring ODE models are emphasized.

- Module 4 (Freshman Spring): **Multivariable Calculus, Part 1.** A first treatment of multivariable calculus, giving a streamlined introduction to the important ideas and techniques in elementary contexts. Vectors, lines, and planes; functions of several variables, partial differentiation and the chain rule, gradient, directional derivative, double and triple integrals, vector fields, the line integral, the surface integral for a box or a sphere, Green's Theorem, Gauss's Theorem and Stokes's Theorem for simple surfaces.

- Module 5 (Sophomore Fall): **Multivariable Calculus, Part 2.** Optimization of functions of several variables, Lagrange multipliers, coordinate systems, change of variable in multiple integrals, parametrization of surfaces, the surface integral for a smooth surface, Divergence Theorem and Stokes's Theorem, applications to electromagnetics and fluid mechanics.

- Module 6 (Sophomore Fall): **Probability and Statistics.** An introduction to elementary probability and statistics. Discrete and continuous probability distributions, independence, conditional probability, Bayes' Theorem, random variables, expectation, variance, joint distributions and independent random variables, covariance, laws of large numbers and Central Limit Theorem; point estimation, hypothesis testing, mean of a normal population, interval estimation, simple linear regression.

- Module 7 (Sophomore Spring): **Linear Algebra, Part 2.** Return to linear algebra; review elementary matrix algebra. Vector spaces and linear transformations, function spaces, basis and dimension, change of coordinates, matrices of linear transformations, change of basis, eigenvalues and eigenvectors, diagonalization, inner products, projection, orthogonalization, least squares, symmetric matrices, quadratic forms. Elementary applications to Markov chains and Fourier series.

- Module 8 (Sophomore Spring): **Differential Equations, Part 2.** Review of solution techniques from Part 1. ODE theory, fundamental solutions, Existence and uniqueness, Wronskian, relationship to linear algebra. Solutions to higher-order equations. Infinite series solutions, Euler equations, regular singular points. Systems of linear equations, stability, the phase plane; almost-linear systems, linearization and local stability analysis, periodic solutions. Applications to physical, engineering, and biological problems. The modular structure provides many opportunities to revisit mathematical ideas and applications with an increasing sophistication. This reiteration of ideas is designed to reduce the compartmentalization of the mathematics, and to explore more connections between different areas of math and applied fields.

Use of Technology

To assist the students, HMC built a large number of Web-based tutorials on topics from calculus, linear algebra, and differential equations. These tutorials focus on key concepts and techniques; many of them have interactive components that allow the student to explore. Each tutorial has several quizzes available, and at any time a student can take a quiz and learn the results immediately. Quiz results are reported to the instructor, and quizzes can be used for review or as assignments in a course.

Cooperative Half-Courses

The mathematics faculty is working with other departments to offer half-course sequences on advanced topics. The first half-course is delivered by a mathematics instructor; and the companion half-course, which finishes the semester, is delivered by an instructor in a science or engineering department. For example, HMC has developed a half-course on **Mathematical Modeling in Biology**, taught by a mathematician; the companion course is on **Biostatistics**, taught by a biologist. Students see a variety of models in the first half of the semester, and they learn how to analyze data and test model hypotheses in the second half. Faculty teams are pairing a mathematical introduction to wavelet theory with a course taught by an engineer on applications of wavelets to signal processing.

Other half-courses planned include: **Variational Calculus** (Physics, Engineering), **Complex Dynamics** (Physics), **Cryptography**, **The Shape of Space** (Physics), **Group Representation** (Chemistry, Physics), and **Mathematical Biology** (Biology). By means of these courses, HMC hopes to improve cooperation between mathematicians and other technical departments and to further emphasize to students the interdisciplinary role that mathematics can play.

Macalester College

Macalester is a liberal arts college that looks for opportunities to capitalize on the already close connections between faculty in Mathematics and other departments. This has resulted in interdisciplinary projects with colleagues in Chemistry, Economics, and Physics. It has also led to a curriculum in Computational Science that draws on interest in other departments in problems such as:

- Chemistry: chemical reaction dynamics using large-scale computing techniques.

- Physics: mathematical methods of classical and quantum mechanics.

- Geography: geographic information systems.

- Sociology: analysis of large sets of demographic data.

- Economics: game theory.

- English: text searching and analysis.

- Biology: finite-element models of cardiac conduction.

- Environmental Science: visual analysis of large data sets.

Computational Science

Computational science is a rapidly growing area in the physical sciences, natural sciences, economics, and engineering. It is a relatively new discipline that investigates how to use mathematical models and high performance computing to attack complex problems drawn from a wide variety of fields. This emerging discipline unites concepts and technologies from both mathematics (modeling, statistics, numerical analysis) and computer science (algorithms, data structures, symbolic computation, scientific visualization).

Computational Science has established itself as the third basic method of investigation of scientific phenomena, alongside theory and experimentation. The reason for the growing interest in computational science becomes clear after looking at some of the massive programs being posed by the High Performance Computing and Communications Initiative. These "grand challenges" are problems in science and engineering that are so large that their solutions are not amenable to experimental techniques. Instead, they can be addressed only by using modeling and high-performance computing. Without this new technique, it would be difficult, if not impossible, to solve a number of very important "big science" problems, such as evolution of galaxies, elementary particle analysis, models of cancerous tissue growth, airfoil design, molecular dynamics, climatic change and long-term weather forecasting, seismic data analysis, and monetary models.

The main thrust of Macalester's project is to develop a sequence of courses in computational science and to support them with a database of computational science topics modules and interdisciplinary application projects.

Courses

Macalester has developed a sequence of two courses suitable for all mathematics and science students. It draws its examples from a range of disciplines, and it focuses on issues and concerns that are broadly applicable to many fields.

The first course is Introduction to Scientific Programming, a course in scientific programming for math and science majors. The course introduces a programming language. The rest of the course includes topics in scientific programming like splines, simulation, optimization. The second course is Computational Science, a modernized numerical analysis course. Macalester is developing a national model for an emerging discipline, one that by its very nature is highly interdisciplinary.

Introduction to Scientific Programming

This course assumes no previous experience with computer programming. It is organized around seven major methodological themes of scientific programming:

- Visualization. Graphical presentation of information in 2 and 3 dimensions, the use of color, animation.

- Optimization. Fitting functions to data, energy minimization in physical systems.

- Databases. Structuring data for effective retrieval, updating, and searching.

- Data Acquisition. Analog-to-digital conversion, scanning, filtering.

- Searching and Comparison. Processing of large datasets, pattern and homology matching.

- Monte Carlo Simulation. Diffusion, ecological models, bootstrapping, and statistical inference.

- Inversion and Transformation. Image processing, principal components.

This course has been offered with several different languages, including Java and Mathematica. It is currently being taught with MATLAB.

Computational Science

This follow-up course takes a deeper look at the ideas introduced in Introduction to Scientific Programming, using a case study approach that presents interesting scientific applications, constructs a model of the problem, solves the model, and then analyzes the output. Among the topics addressed in this course are computational modeling, advanced scientific visualization, scientific libraries and toolkits, and parallel and high-performance computing.

Texas Southern University

The college-algebra reform project led by Texas Southern began with a consortium of seven Historically Black Colleges and Universities (HBCUs). The consortium has grown to 14 schools. College algebra has extremely high enrollments at many schools because it often is the first course in the mathematics sequence to receive college graduation credit and is required by many disciplines of study. At several schools, its enrollments exceed the enrollments of all the other mathematics courses combined. Unfortunately, the traditional college algebra course frequently acts as a "gate-keeper" and "filter" that deflects students into nonquantitative studies or prevents further undergraduate study. These courses usually contain little or no use of technology and limited use of small-group work and cooperative learning. There is no emphasis on communication skills, interdisciplinary applications, or problem solving.

Goals of the Reform Program

The reformed program has two major goals. The first is to develop a new course that will

- be activity and project-based—small group activities and projects developed by interdisciplinary teams of faculty;

- incorporate a strong technology component;

- develop problem solving skills;

- develop communication skills;

- provide a conceptual understanding of algebra and develop skills in using algebraic techniques; and

- develop the students' mathematical self-esteem and confidence as problem-solvers.

The second goal is to change the culture surrounding the traditional college algebra program by

- developing a sense of ownership and pride in the program among the teaching faculty;

- energizing faculty to develop modes of instruction that actively engage students in their own learning; and

- developing a sense of involvement and responsibility for the program on the part of faculty in disciplines that require college algebra for their majors.

New Course

The new course is designed to aid the student in developing "mathematical life skills" to answer three categorical questions:

- How to present data?

- How to transform data into information?

- How to transform information into knowledge?

The course was developed in consultation with faculty in partner disciplines. These faculty members strongly requested elementary data analysis, plotting and interpreting graphs, and modeling real situations, including such activities as collecting data on vehicle stopping times, plotting the data, fitting a curve to the data, and then using the resulting function for predictive purposes.

Two or three Fun Projects are assigned during the semester course. These are small-group, out-of-class projects that culminate in a written report. Most of the projects are inquiry-based, requiring students to research a situation, conduct interviews, collect data, and so on. Faculty development workshops are designed to assist instructors develop small-group activities and projects and to teach with technology. This project is transforming the college algebra "filter" into a "pump" to change significantly the culture of mathematics on many HBCU campuses.

The consortium faculty meet for a professional development retreat in the fall, during the Joint Mathematics Meetings, and in the spring for a workshop. In addition, the consortium publishes a newsletter (eight times a year) and holds various dissemination workshops.

New Text

The new textbook (Small [2001]) contains student-relevant applications and motivational examples and exercises. There are three sections in the text:

- "Data and Variables" includes elementary data analysis and linear relations.

- "Functions" is motivated by the need to form predictions based on data and includes graphical and symbolic approximations of data.

- The third section focuses on modeling in several different disciplines, using recursive sequences.

Each section concludes with a description of six to eight small-group projects.

Response

Both student and faculty attitudes have changed from negative to positive, even though more time and effort are required for both groups. In particular, students show a great deal of pride in their project reports. Pass rates have increased and dropout rates have declined markedly. It is not uncommon for an instructor to report that no student dropped his or her class. Faculty discuss, with pride, what their students can do, rather than what they cannot do.

References

COMAP. 1996. *Principles and Practice of Mathematics*. New York: Springer-Verlag.

Small, Don. 2001. *Contemporary College Algebra*. New York: McGraw-Hill.

How to Develop an ILAP

Joseph D. Myers
Dept. of Mathematical Sciences
United States Military Academy
West Point, NY 10996
aj5831@usma.edu

Introduction

In this guide, we briefly explain what an Interdisciplinary Lively Application Project (referred to hereafter as an ILAP) is, how they are developed and executed, and some considerations and strategies in developing and using ILAPs. We include only the essentials. There are many perspectives and elements to consider, but we leave those for future articles.

ILAP Is a Process That Generates a Product That Drives a Student Learning Experience

ILAPs are student group projects that are jointly authored by a faculty member from the Mathematics Department and a faculty member from a partner department. ILAPs can be used in the mathematics classroom, in the partner classroom, or in both to let students work on mathematical concepts within the context of another discipline.

Characteristics of ILAPs:

- Authored jointly by faculty from partner disciplines

- Connected to the curricula of both partners, and can reach forward or backward to connect current topics with ideas already studied or to preview ideas that wait downstream.

ILAPs incorporate and provide students practice in the interdisciplinary threads of:

- modeling,

- reasoning (within an applied context!) and problem solving,

- using technology as a tool to enable analysis,

- connecting and integrating,

- teamwork, and

- communication.

The product format can be whatever you want, but experience shows that the following elements are useful:

- Problem statement.

- Background material. This is often as important for instructors as it is for students, since the application discipline is often outside their expertise.

- Sample solution. Not intended to be used for grading, but rather to boost the instructors' comfort level and understanding of the problem.

- Report/briefing guidance for student authors/presenters.

Benefits sought through ILAPs:

- Motivation through relevance. Show students how current mathematical ideas are used in other disciplines, and demonstrate the areas of learning that become accessible to them as they learn and master mathematics.

- Support student growth in the interdisciplinary threads listed above.

- Give students experience in solving problems as part of a team.

- Develop partnerships to discuss and develop curricula.

     After you have used ILAPs for a while, you will realize that

the most valuable part of ILAPs is the process of teaming and working with a partner department, rather than the product that is created at the end.

We have authored many ILAPs at USMA. We initially thought that we needed only to develop a few that we could then refine and recycle on a several-year cycle. However, we typically use each ILAP only once or twice and spend our time partnering and developing new ILAPs rather than refining and reusing and publishing old ones.

Developing and Executing ILAPs

The usual chronology for developing an ILAP is:

- **Decide on a topic and skills** to be used; these are usually current topics and skills from the mathematics curriculum.

- **Approach a faculty member in another department** with what you want to do, and ask for their ideas on an application that uses these skills. This is the part that involves some teamwork, communication, and creativity on the part of the faculty—the same habits we seek to develop in our students! One of the important goals of the ILAP process is to gradually develop an interdisciplinary culture where partner faculty take the initiative in seeking out mathematics faculty to develop ILAPs that cover concepts and skills needed in the partner disciplines' courses. It is unlikely that this will happen immediately; however, your willingness to begin and nurture the process is the one key factor in making it happen.

- **Write the ILAP jointly** with the partner department. Incorporate the mathematical topics and skills that you want exercised, and include the partner scenarios, ideas, and connections that will be seen again by students in the partner discipline. This can be either a true joint process where both authors sit together, or an iterative process where one author (from either discipline) begins sketching the product and then the product is refined through a series of iterations/input/revisions between authors. Distribute the student handout and make the initial presentation to students. This usually involves prompting the students for what they already know about the scenario, helping them identify assumptions, seeing if they have some initial guesses about what a reasonable resolution would look like, and so on. This initial presentation is often made by the mathematics faculty but can be made with great effect by the partner faculty. It is a powerful message to students when the partner faculty visits the mathematics classroom, explains the scenario from their discipline, and challenges the students to learn the mathematics that will enable them to begin to succeed in the partner discipline.

- **Have student groups work on the ILAP.** An effective technique is to not work on portions of the ILAP in the intervening classes, but rather to ask students how they think the material they are currently working on in class could apply and be used in the ongoing ILAP. However, be careful not to let students "divide-the-work and submit" or wait until the last minute to begin working on the project itself.

- **Have student groups make written and/or oral presentations** of their analyses and solutions. Written presentations are an excellent opportunity to develop skills in writing technical reports, with executive summaries, assumptions, analysis, conclusions, and supporting technical appendices. Students appreciate regular oral presentations as well; it gives them the opportunity to express themselves in person, instructors can ask clarifying questions on the spot rather than being forced to rule on written vagaries, and team members usually work together in a more integrated and balanced fashion.

- **Have the partner faculty give an expert critique** and extension of student submissions/presentations. This is effective in bringing closure to the project, in showing how the basic ideas and skills just used are amplified and made more sophisticated in the downstream discipline, and in reinforcing that the students have prepared themselves to study and understand another discipline.

Strategy for Using ILAPs

Experience has shown us that ILAPs can be made more useful and successful by authors' consciously using the following strategies:

- Be sufficiently understandable and comfortable for faculty (but it's all right and even healthy to get a bit outside of faculty expertise—it reminds us of what it's like to be a student).

- Involve the students—keep things accessible at their level so they retain responsibility.

- Capitalize on students' intuition and increase their ability to verify it.

- Incorporate technology.

- Increase and intensify faculty cooperation.

- Be flexible, multifaceted, and open-ended.

- Contain recommended guidance for students who need it.

- Admit the need for (or require):

    - Research
    - Discovery
    - Teamwork
    - Individual responsibility

- Focus on one or a few basic mathematical concepts.

- Identify connections among the applications and disciplines.

- Encourage discussions and properly cited collaborative work between groups and others.

- Encourage multiple and ingenious approaches through graphical, numerical, and analytical techniques.

- Be self-contained but able to be revisited in the partner discipline.

Considerations in Developing an ILAP

After developing many ILAPs for different levels of students in a variety of courses and with many different partner departments, we have found the following mental checklist of considerations to be useful during development:

- What mathematical **topic and skills?**

    - A mathematical topic that needs some life/application?
    - Mathematical skills from another discipline that students struggle with?

- What **discipline/application** to connect with?

    - Play to student interest/intuition?
    - Mathematics faculty comfort/interest?
    - Partner faculty interest? Potential for revisiting?

- **Scenario?** (Again, possibly playing to student interest or intuition)

- **Sophistication:**

    - Modeling: How much provided, how much done by the student?
    - More prescriptive, more open-ended, or both?

- **Range of difficulty** in requirements:

    - Easier? (analytic by hand)
    - Harder? (numerical, visual, analytic with technology)

- **Responsibility:**

    - Requirements unique to each student group? It is valuable to have all groups working on the same scenario and requirements; students report learning from others when all groups have a common task. But to encourage group accountability and ownership, you may want to give each group a unique set of parameter values. We have done this for ILAPs in courses of up to 1,000 students and have been pleased with the balance between collaborative learning and group accountability and ownership.

    - Individual accountability of team members? Within assigned groups, one wants to see a balance between collaborative effort and individual contribution to the project. Common difficulties within assigned groups are to see one or two members do all the work and others do little, or to see a "divide-the-work and submit" approach in which each member does a portion in isolation and the results are stapled together and submitted with no collaboration and no member having thought about the big picture of the project as a whole. Options are to

 * give a follow-up quiz on the project,
 * put a simplified project requirement on the midterm or final, or
 * opt for oral student presentations and spread questions among all group members.

Of course, advertising to students that they will be accountable in these ways is the key to encouraging individual participation within each group.

• **Presentation:**

 – Written and/or oral?

 – Expert (partner) critique, summary, extension?

• **Time commitments:**

 – How much time per student? It is very important to know about how much student time an ILAP will require. Experience and instructor solutions help here. We figure on a ratio of

   1 instructor hour = 3 student hours.

We have found that projects work well when 8 or 9 hours are required from each student.

 – Designate time in syllabus? Students need quality time to do a quality job; rather than poach on their time, we build into the syllabus the expected amount of student time required. This is usually done with class drops or reduced assignments.

• **Grading:**

 – Instructor expectations? Know what results you expect before you assign a project. Different levels of project difficulty can be accommodated in how you grade, as long as you know what you expect of your students.

 – Cut sheet? Some faculty members find it useful to prepare a sheet of what they expect from students' final products. Grading against this type of cut sheet explicitly tells students what is expected and valued and can help standardize the grading in a large course. Others prefer to grade against their mental expectations; doing so is certainly quicker and allows for more judgment, including adjusting expectations based on what most students are doing,

Guidance for Students on Written Reports

A written report is a great way for students to communicate what they did and have found in their projects. However, effective report writing is not easy, and students need help in developing their skills through practice, experience, and feedback. As students progress through their programs, most will be required to do technical writing in other courses. To the extent that faculty can coordinate their technical writing needs and expectations for students, we can help students grow and mature as problem solvers who can effectively communicate their analyses and recommendations to others for adoption and action.

There is no best format guide, but as an example we summarize the writing guidance that we give our students. This format fits our needs and is consistent with the format used in the courses of our partners.

- **Executive Summary:** A one- or two-page summary of the scenario, what questions were addressed, how they were addressed, and -the students' conclusions and recommendations-. The summary sometimes can be formatted as a letter addressed to the client or user in the scenario, if appropriate.

- **Problem Statement:** This is a concise summary of the issues of interest in the given scenario.

- **Facts Bearing on the Problem:** These are facts either known from the problem statement or uncovered during the research and problem-solving stages.

- **Assumptions:** These fill the gaps between what is known and what is required to do a successful analysis. Each must be necessary, not provable from known facts, justified as to why one should believe that each might reasonably be true, and if possible checked for consistency at the end of the analysis.

- **Analysis:** This represents the heart of the work. It can include the following sections as appropriate:

  - Definition of Variables and Symbols,
  - Methodology Used,
  - Formulas Used,
  - Calculations,
  - Essential Graphs and Diagrams, and
  - Discussion of Results.

Often, many of these areas are combined into the analysis narrative rather than separated into sections. It is critical that the Analysis section be presented in narrative form, with equations and graphs used to clarify the exposition; students are not allowed to present long multi-line derivations with no explanatory text. Long derivations or supporting work that disrupts the narrative is only referenced here and is presented instead in an appendix (see below).

- **Conclusions and Recommendations:** These must follow logically from the analysis narrative. No new material should be introduced here. The presentation here must directly address the issues of interest from the scenario.

- **Appendices:** Long derivations and supporting work that disrupts the analysis narrative is presented here and referenced from the narrative. Each appendix must be cited somewhere in the main body.

- **Acknowledgments:** All sources from outside the group are to be given due credit here. We expect some amount of learning to happen from other groups, but each group must give specific credit to the group, paper, or other source from which they receive assistance. Acknowledgments also must be specific regarding what assistance was received from that source. Normal and healthy assistance is encouraged and is not penalized; excessive reliance on a single source in a given part of the project is discouraged and may result in a lower academic grade.

Grading

As our instructors grade reports, they use the following institutional standards for writing. Substance is the key area that we grade on, but instructors make corrections from all four areas on student projects and can and do make grade adjustments in all areas for strengths or weaknesses that they note.

You may want to incorporate similar standards from your institution in your grading process. Our four writing areas are:

- **Substance:** The correctness, completeness, and persuasiveness of the exposition.

- **Organization:** The logical flow of the report. This format guide, or a suitable student variant, is usually a big help in this area.

- **Style:** Avoiding slang, undefined acronyms, undefined or inappropriate technical jargon, and excessive use of the passive voice. Style is important in technical writing; achieving the right balance between discourse and technical expression is key to communicating with the reader.

- **Correctness:** The document must be free of spelling, grammatical, and punctuation errors. The correctness of the mathematics and logic is considered under the **Substance**.

Developing Students

We have an obligation to our students to:

- provide them experience working in small groups;

- develop skill in the use of technology;

- develop communications skills (reading, writing, and presenting);

- develop self-esteem and confidence as problem-solvers; and

- demonstrate ways that their learning connects to the rest of their curriculum.

We have found ILAPs to be a great way to provide appropriate developmental experiences in these areas.

About the Author

Joe Myers graduated from USMA in 1978 and holds a degree in Applied Mathematics from Harvard. He has taught at USMA since 1987, and has served as an Academy Professor there since 1991. He enjoys helping students connect the various pieces of their learning and discover that the intersection of many of their fields of study lies within mathematics.

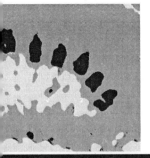

INTERDISCIPLINARY LIVELY APPLICATIONS PROJECT

UTHORS:
lye I. Selco (Chemistry)
lye_selco@redlands.edu

net L. Beery (Mathematics)
net_beery@redlands.edu

iversity of Redlands
dlands, CA

ITOR:
vid C. Arney

Saving a Drug Poisoning Victim

MATHEMATICS CLASSIFICATIONS:
Calculus, Differential Equations, Mathematical Modeling

DISCIPLINARY CLASSIFICATIONS:
Chemistry, Biology, and Medicine

PREREQUISITE SKILLS:
Exponential growth and decay, Euler's method or other numerical
method for solving systems of differential equations

PHYSICAL CONCEPTS EXAMINED:
Kinetics of drug uptake and elimination

MATERIALS INCLUDED:
TrueBASIC programs

COMPUTING REQUIREMENTS:
Numerical differential equations solver, spreadsheet, computer
algebra system, or any computer programming language

Contents

1.   Setting the Scene

You are a physician in a hospital emergency room. A child has just been brought to the emergency room by a frantic parent. The parent takes the asthma medication theophylline in tablet form. Two hours before arriving at the hospital, the child ingested eleven 100-mg theophylline tablets. Like most oral drugs, theophylline is absorbed into the bloodstream at a rate proportional to the amount present in the gastrointestinal tract (stomach and intestines) and is eliminated from the bloodstream at a rate proportional to the amount present in the bloodstream.

Your quick check of the *Physician's Desk Reference (PDR)* [1999] reveals that the brand of theophylline that the child took has an absorption half-life of 5 hours and an elimination half-life of 6 hours. The *PDR* also warns that a blood-level concentration of 100 mg/L or more of the drug is seriously toxic and that a concentration of 200 mg/L or more is fatal.[1]

You estimate that the child has 2 L of blood. You also determine that because of the 2-hour delay, the pills already have passed from the child's stomach to his intestines, so that it is too late to eliminate the drug by inducing vomiting. Your task is to determine if the child is in danger, and, if so, to save his life.[2]

2.   Building a Model

You are interested in the amount of theophylline in the child's bloodstream over time. (Actually, you are concerned about the *concentration* of theophylline in the child's bloodstream over time; but since the amount is slightly easier to

---

[1]These values are the concentrations at which 50% of the patients exhibit these symptoms. In the fatal case, the concentration of of 200 mg/L—the lethal concentration for 50% of the population—is called the LC_{50} value.

[2]In reality, a physician in this situation would contact the local poison center, which would provide information about which symptoms to watch for as well as the appropriate medical treatment.

Figure 1. Compartment model for theophylline.

calculate than the concentration and since you can convert easily from one to the other, you decide to calculate the amount.)

To determine the amount over time, you also need to determine the amount of theophylline still in the child's gastrointestinal tract over time. You could calculate also the amount of theophylline eliminated from the bloodstream; however, since theophylline in this form is not dangerous, you decide not to keep track of the eliminated drug. The compartment model in **Figure 1** illustrates the progress of the drug through the child's body.

Requirement 1: First, predict the general shape of the graph of $G(t)$, the amount of theophylline in the child's gastrointestinal tract (in mg) after t (in hours), and of the graph of $B(t)$, the amount of theophylline in the child's bloodstream (in mg) after t hours. Using time $t = 0$ as the time at which the child first ingested the theophylline, make separate rough sketches of the graphs of $G(t)$ and $B(t)$. On each graph, label the point at $t = 0$. (If $t = 0$ is the time when the child first ingested the theophylline, what are the corresponding values for G and B?) Remembering that the half-life for absorption of theophylline from the gastrointestinal tract into the bloodstream is 5 hours, label the points at $t = 5$ and $t = 10$ on your graph of $G(t)$. You need not label any other points on the graphs or mark any other values along their axes—yet.

Requirement 2: Since you have more information about the rates of change of G and B than about G and B themselves, you decide to model the quantities G and B by writing equations for their rates of change (differential equations). Begin with the differential equation for G. Theophylline is absorbed into the bloodstream at a rate proportional to the amount present in the gastrointestinal tract. This means that theophylline is *leaving* the gastrointestinal tract at a rate proportional to the amount of the drug present there. Hence, taking k to be the positive constant of proportionality, you have

$$\frac{dG}{dt} = -kG \text{ mg/h}, \qquad G(0) = 1100 \text{ mg}.$$

Use what you know about initial value problems of this form, along with the fact that the absorption half-life of theophylline is 5 hours, to write a formula for $G(t)$, the amount of theophylline (in mg) in the gastrointestinal tract at time t. (That is, solve the initial value problem for $G(t)$, then solve for k. Record k to four decimal places.) You now should have both a formula for $G(t)$ and a differential equation for G in which k has a numerical value.

Requirement 3: Now write a differential equation for B. Since theophylline is entering the bloodstream at one rate and leaving it at another rate, the differential equation for B is of the form

$$\frac{dB}{dt} = \text{absorption rate} - \text{elimination rate},$$

with units of mg/h.

Consider the first term, the absorption rate. Recall that theophylline is absorbed into the bloodstream at a rate proportional to the amount present in the gastrointestinal tract with an absorption half-life of 5 hours. This should sound familiar; use your work from **Requirement 2** above to write an expression for the absorption rate.

Now consider the second term, the elimination rate. Remember that theophylline is eliminated from the bloodstream at a rate proportional to the amount present in the bloodstream with a half-life of 6 hours. In order to find the constant of proportionality, assume that at some (future) time t_1 there is 20 mg of theophylline in the bloodstream and that no additional theophylline is entering the bloodstream—that is, assume for the moment that

$$\frac{dB}{dt} = - \text{elimination rate}, \qquad B_1(t) = 20 \text{ mg}.$$

Under these assumptions, the amount of theophylline in the bloodstream is decaying exponentially. Use what you know about exponential decay to write an expression for the elimination rate. (Record the constant of proportionality to four places after the decimal point.)

You now should have a differential equation for B involving the variables G and B.

3.   Using the Model

Now that you have differential equations for G and for B, you are ready to use them to determine if the child is in danger and, if so, how to treat him.

Unlike for the differential equation for G, there is not a simple closed-form solution for the differential equation for B. That is, you may not be able to write an explicit formula for $B(t)$ but instead may have to approximate values of $B(t)$ using Euler's method or another numerical method for solving differential equations. Your instructor will specify the degree of accuracy (number of significant figures) for your calculations.

Requirement 4: Determine the amount of theophylline in the child's bloodstream at the time of his admission to the hospital, $t = 2$ hours. Recalling that the child has 2 L of blood and that a blood-level concentration of 200 mg/L or more of the drug is fatal, what amount of theophylline in his bloodstream, in mg, constitutes a lethal level for the child? Recalling that a blood-level concentration of 100 mg/L or more is seriously toxic, what amount constitutes a seriously toxic level for the child? What is his status at the time of his admission to the hospital? Does the amount of theophylline in his bloodstream pose any danger to him at this time?

Determine the toxic and lethal amounts of theophylline in the bloodstream for an adult with 6 L of blood.

Requirement 5: Determine the amount of theophylline in the child's bloodstream over several hours. Graph your results.

Does the amount of theophylline in his bloodstream ever reach a lethal level for the child? If so, after how many hours? How many hours after the child's hospital admission does this occur? Mark the lethal level and the time at which it occurs on your graph.

After how many hours does the amount of theophylline in the child's blood reach a seriously toxic level? How long before or after his hospital admission does this occur?

Requirement 6: Determine the largest amount of theophylline the child ever has in his bloodstream, and the time at which this maximum level occurs. How much theophylline remains in his gastrointestinal tract at this time?

The largest value for B occurs when $dB/dt = 0$ (why?) or, equivalently, when the absorption rate is equal to the elimination rate (why?). When you substitute your largest value for B and your corresponding value for G into your equation for dB/dt, do you get 0? Explain why you might not get 0.

Requirement 7: Determine the maximum number of theophylline tablets the child could have taken in a short interval without reaching the lethal blood-level concentration. Explain. How many could he have taken without reaching the seriously toxic blood concentration? Explain.

4. Saving the Child

Your results from **Requirements 5–7** should have shown that the child who ingested the 11 theophylline tablets is in grave danger. What can be done?

Fortunately, charcoal absorbs theophylline quickly, so it can be used to increase the rate at which theophylline is eliminated from the bloodstream. For toxic levels of theophylline, the patient takes oral doses of charcoal, increasing the theophylline elimination rate to approximately twice the normal rate.

For potentially fatal levels of theophylline, charcoal must be added to the bloodstream extracorporeally (outside the body) in order to remove the theophylline quickly enough. This procedure is risky but may increase the theophylline elimination rate to six times the normal rate, according to the *Physicians*

Desk Reference [1999].[3]

Requirement 8: In **Requirement 3**, you expressed the rate of change of B as the difference between the absorption rate and the elimination rate. Since it is too late to change the rate of absorption from the gastrointestinal tract, you must change the elimination rate, which you have expressed as $-cB$ mg/h. Find the smallest value for the constant c that ensures that the concentration of theophylline in the child's bloodstream remains below the lethal level. For instance, would increasing c to 0.1200 suffice? What about $c = 0.1300$?

Warning: Since you have an opportunity to increase the value of c only after the child has been admitted to the hospital, be sure to begin your calculations with larger values of c at the time of admission.

To be safe, continue to increase c until you find the smallest value that causes the amount of theophylline in the child's bloodstream to decrease immediately upon treatment. Assume that treatment is administered exactly at the time of hospital admission. Sketch or print a graph of the amount of theophylline in the child's bloodstream from the time he ingests the pills to a few hours after his hospital admission and treatment. Your graph should show the effect of treatment on the child's theophylline blood level.

Without any further computer work, you could have determined the smallest value for c that causes the amount of theophylline in the child's bloodstream to decrease immediately upon treatment. Explain how. (Hint: See **Requirement 6**, second paragraph.)

Recall that you could have doubled the theophylline elimination rate by administering oral doses of charcoal. Would this treatment have been sufficient to cause an immediate decrease in the child's theophylline blood level? Explain. Remember that you can increase the elimination rate to six times the normal rate by filtering the blood through charcoal extracorporeally. Do you need to increase the elimination rate this much in order to cause an immediate decrease in the child's theophylline blood level? Explain.

References

The problem presented in the **Setting the Scene** section appeared in slightly different form in a physical chemistry textbook [Bromberg 1984, 905]. The tablet size and absorption and elimination half-lives appeared in the original problem statement and are consistent with the *Physician's Desk Reference* [1999] listings for various brands of theophylline tablets. Treatment methods for theophylline overdose are described in both the *Physician's Desk Reference* and the *Physician's Desk Reference Generics* [1998]. Some of the questions posed above were inspired by the development of the susceptible-infected-recovered (S-I-R) model for the

---

[3]Since the lethal concentration represents the response of 50% of the population and dialysis is both a dangerous and expensive, the physicians would resort to such drastic treatment only if the patient would not survive without it.

spread of an epidemic in the textbook *Calculus in Context* [Callahan et al. 1995, 1–69]. In particular, the True BASIC computer programs provided to your instructor were adapted from that textbook. The graphs provided to your instructor were created using the Excel spreadsheet program.

Bromberg, J. Philip. 1984. *Physical Chemistry*. 2nd ed. Boston: Allyn and Bacon.

Callahan, James, Kenneth Hoffman, et al. 1995. *Calculus in Context*. New York: W.H. Freeman.

Physician's Desk Reference. 1999. 53rd ed. Montvale, NJ: Medical Economics Co. See, for instance, pp. 875–878, 1457–1471, 2563–2569, 2607–2609, and 3164–3170.

Physician's Desk Reference Generics. 1998. 4th ed. Montvale, NJ: Medical Economics Co. See pp 2620–2633, especially pages 2626–2627.

Acknowledgments

The authors thank the staff at Redlands Community Hospital and the Southern California Poison Center for medical information.

Title: **Saving a Drug Poisoning Victim**

Sample Solution

Requirement 1: Although the graphs of $G(t)$ and of $B(t)$ should be sketched by hand on separate sets of axes, they should have essentially the shapes shown in **Figure S1**. The points $(0, 1100)$, $(5, 550)$, and $(10, 275)$ should be labeled on the graph of $G(t)$; the point $(0, 0)$ should be marked on the graph of $B(t)$. Students will have the opportunity to verify their predictions after they set up differential equations for G and for B.

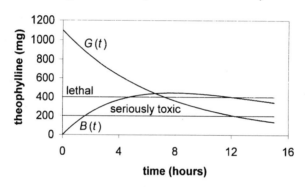

Drug Overdose (without treatment)

Figure S1. Graphs of $G(t)$ and $B(t)$.

Requirement 2: Students should recognize the differential equation for exponential decay and write $G(t) = 1100e^{-kt}$ mg. They then should use that $G(5) = 550$ mg to solve for k, obtaining $k = (\ln 2)/5 \approx 0.1386$. The formula for $G(t)$ then is $G(t) = 1100e^{-0.1386t}$ mg. The initial value problem for $G(t)$ is

$$\frac{dG}{dt} = -0.1386G \text{ mg/h}, \qquad G(0) = 1100 \text{ mg}.$$

Requirement 3: The absorption rate is given by the expression $0.1386G$ mg/h from **Requirement 2**. To find an expression for the elimination rate, we assume that

$$\frac{dB}{dt} = -cB \text{ mg/h}, \qquad B(t_1) = 20 \text{ mg},$$

yielding $B(t) = 20e^{-ct}$ mg. Students then should use $B(t_1 + 6) = 10$ mg to solve for c, obtaining $c = (\ln 2)/6 \approx 0.1155$. The elimination rate then is given by the expression $-0.1155B$ mg/h, and the initial value problem for $B(t)$ is

$$\frac{dB}{dt} = 0.1386G - 0.1155B \text{ mg/h}, \qquad B(0) = 0 \text{ mg}.$$

Students may conjecture a solution to this initial value problem of the form

$$B(t) = \mp 1100e^{-0.1386t} \pm B_0 e^{-0.1155t}$$

and should be encouraged to check to see that this is not a correct solution. A correct solution, obtained by matrix methods, is

$$B(t) = -6600e^{-0.1386t} + 6600e^{-0.1155t}.$$

Although this project could be modified to incorporate this analytic solution, we assume that students will solve the differential equation(s) numerically.

Requirement 4: Answers are given to 4 significant figures. (Please see the note on **Accuracy of Solutions** in the **Notes for the Instructor**.)

Applying Euler's method with a step size of $\Delta t = 0.0001$ to the initial value problem

$$\frac{dG}{dt} = -0.1386G \text{ mg/h}, \qquad\qquad G(0) = 1100 \text{ mg},$$

$$\frac{dB}{dt} = 0.1386G - 0.1155B \text{ mg/h}, \qquad B(0) = 0 \text{ mg}$$

yields $G(2) = 833.7$ mg and $B(2) = 236.5$ mg. Computing $G(2)$ using the formula for $G(t)$ yields $G(2) = 833.7$ mg. One also could use the formula for $G(t)$ to compute the values of G needed in the equation for dB/dt during the Euler's method computations.

The VALUE program listed in the **Appendix** is set up to approximate $G(2)$ and $B(2)$ using Euler's method with a step size of $\Delta t = 0.1$ (20 steps) and can be modified to perform the calculation with $\Delta t = 0.0001$ (20,000 steps).

Since the child has 2 L of blood, a lethal blood level of theophylline for him would be 400 mg, well above his current 236.5-mg blood level. However, only 200 mg would constitute a seriously toxic blood level for him, and his current 236.5-mg blood level already is in the seriously toxic range.

For the adult with 6 L of blood, 600 mg would be seriously toxic, while 1200 mg would be fatal.

Requirement 5: By computing values of B from $t = 0$ to approximately $t = 10$, students can see that the amount of theophylline in the bloodstream increases and then decreases, and that it does eventually exceed the lethal level (**Figure S1**). The amount of theophylline in the bloodstream reaches $B = 400.0$ mg after $t = 4.866$ h (2.866 h after the child's admission to the hospital), and $B = 200.0$ mg after $t = 1.609$ h (approximately 23 min, 28 sec before the child's hospital admission).

By changing tfinal to 10 or more hours in the VALUE program, students can see that B increases and then decreases, and that it does eventually exceed 400 mg. Running the PLOT program illustrates this behavior even more clearly. To determine how many hours it takes for B to reach the lethal level of 400 mg, students may use the DOWHILE program. Running the DOWHILE program with $\Delta t = 0.0001$ yields $B = 400.0$ mg after $t = 4.866$ h, and $B = 200.0$ mg after $t = 1.609$ h.

Requirement 6: A step size of $\Delta t = 0.0001$ yields $G = 368.4$ mg and $B = 442.1$ mg after $t = 7.893$ h. Note that $B = 442.1$ mg exceeds the child's lethal level of 400 mg.

Substituting $G = 368.4$ and $B = 442.1$ into the equation for dB/dt yields $dB/dt = 0.0023$, indicating that our solution technique is approximate rather than exact and/or that round-off error has occurred. Students may use the formula for $G(t)$ to obtain the value of G at the time at which the largest amount of theophylline in the child's bloodstream occurs or to check the accuracy of their numerical approximations. Computing $G(7.893)$ using the formula for $G(t)$ yields $G = 368.4$ mg.

Requirement 7: If the child had taken only 9 tablets ($G(0) = 900$ mg), the largest amount of theophylline in his bloodstream would have been 361.7 mg ($\Delta t = 0.0001$), which is below the lethal level. If he had taken only 4 tablets, his highest blood level of the drug would have been 160.8 mg; 10 tablets and 5 tablets, respectively, would have resulted in maximum blood levels of just over 400 mg and 200 mg. These computations can be made by changing the initial value for G in the DOWHILE program.

Requirement 8: Starting at the time of hospital admission ($t = 2$ h), students should increase c. They may need to use that $G(2) = 833.7$ mg and $B(2) = 236.5$ mg from **Requirement 4**. Students who did not record $G(2)$ in **Requirement 4** can compute $G(2)$ easily from their formula for $G(t)$.) Using $c = 0.1442$ results in a maximum value for B of $B = 399.9$ mg, whereas using $c = 0.1441$ results in a maximum value for B of $B = 400.0$ mg.

Using $c = 0.4833$ results in a maximum value for B of $B = 236.5$ mg, whereas using $c = 0.4832$ results in a maximum value for B of $B = 236.6$ mg. Or students may notice that using $c = 0.4886$ results in the maximum value for B of $B = 236.5$ mg occurring at time $t = 2.000$ h, whereas using $c = 0.4885$ results in the maximum value for B occurring when $t = 2.001$ h. These results can be obtained by using a step size of $\Delta t = 0.0001$ and initial values of $G = 833.7$ mg and $B = 236.5$ mg in the DOWHILE program.

A graph showing the effect of treatment on the child's theophylline blood level is shown in **Figure S2**.

Note that B decreases when $dB/dt < 0$, or, equivalently, when the elimination rate is greater than the absorption rate. If we set

$$\frac{dB}{dt} = 0.1386G - cB = 0$$

at time $t = 2$, we obtain $(0.1386)(833.7) - c(236.5) = 0$, so that $c = 0.4886$.

Since $2 \times 0.1155 = 0.2310 < 0.4833$, administering oral doses of charcoal would not have been sufficient to cause the child's theophylline blood level to decrease immediately. Since $6 \times 0.1155 = 0.6930$, we did not need to increase the elimination rate by six times the normal rate in order to cause B to decrease immediately. In fact, our value for c is approximately 4.2×0.1155.

Drug Overdose (with treatment)

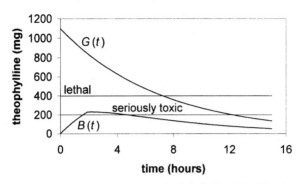

Figure S2. Effect of treatment on the theophylline blood level.

Title: **Saving a Drug Poisoning Victim**

Notes for the Instructor

Suggested Course Use

This ILAP was designed for a calculus course in which students learn Euler's method. It also would be appropriate for a differential equations course or a mathematical modeling course and perhaps even for very ambitious pre-calculus students. Briefly, the ILAP is a differential equations modeling project in which students, posing as hospital emergency room physicians, save a child who has accidentally overdosed on asthma medication. They begin by setting up a system of linear first-order differential equations (DEs) describing the medication's absorption into and elimination from the child's bloodstream. By solving the differential equations numerically, students discover that the child almost certainly will die if they, as physicians, do not intervene. They then determine by how much they need to increase the drug's elimination rate in order to save the child.

The Model

Problems involving drug absorption and elimination appear in many calculus texts, especially "reform" texts, among collections of mixing problems. What distinguishes this problem from any that we have seen in a calculus text is our assumption that the absorption rate is proportional to the amount yet to be absorbed, rather than constant, as it is in standard mixing problems. Specifically, our differential equation for the amount y of the drug in the bloodstream is of the form

$$\frac{dy}{dt} = ax - by,$$

where a and b are constants and x is the amount yet to be absorbed, rather than of the form

$$\frac{dy}{dt} = a - by.$$

The latter DE has a closed-form solution, easily found by separating variables. Assuming the DE for $x(t)$ to be of the same form as that for $y(t)$, a closed-form solution for the former DE can be found using matrix methods. Although this project could be adapted to incorporate analytic rather than numerical solutions, we assume that students will solve the differential equations numerically.

Computing Requirements

To complete the project, students need technology to implement Euler's method or another numerical method for solving systems of differential equations, such as a spreadsheet program, computer algebra system, differential equations solver, or virtually any computer programming language, including that available on a graphing calculator. The project also can be adapted for use with a computer algebra system or other differential equations solver capable of providing analytic solutions to systems of linear first-order differential equations.

One of the authors has her calculus students complete the project by modifying True BASIC programs set up to analyze a Lotka-Volterra predator-prey population model using Euler's method. True BASIC programs set up to analyze this ILAP's drug uptake and elimination model using Euler's method are provided in the **Appendix**.

Accuracy of Solutions

The instructor should specify the number of significant digits to which students are to work, based on such considerations as software speed. We assume that the tablet weights and the absorption and elimination half-lives given in **Setting the Scene** are exact—or at least are accurate to the number of significant digits the instructor specifies. For instance, when we give answers in the **Sample Solution** accurate to 4 significant figures, we assume that the absorption half-life is 5.000 hours.

Title: **Saving a Drug Poisoning Victim**

Appendix: True BASIC Computer Programs

```
!Program VALUE (True BASIC)

!Program uses Euler's method to estimate values of G and B
!tfinal hours from start.

LET tinitial = 0
LET tfinal = 2
PRINT "  t"," G"," B"
LET t = tinitial
LET G = 1100
LET B = 0
PRINT t, G, B
LET Gprime = - 0.1386*G   !Compute G', B' for t = 0.
LET Bprime = 0.1386*G - 0.1155*B
LET numberofsteps = 20
LET deltat = (tfinal-tinitial)/numberofsteps
FOR k = 1 TO numberofsteps
    LET deltaG = Gprime*deltat
    LET deltaB = Bprime*deltat
    LET t = t + deltat
    LET G = G + deltaG
    LET B = B + deltaB
PRINT t, G, B
    LET Gprime = -0.1386*G
    LET Bprime = 0.1386*G - 0.1155*B
NEXT k
!PRINT t, G, B                        !Use for larger numbers of steps.
END


!Program PLOT (True BASIC)

!Program uses Euler's method to plot graphs of G and B together.
!Program also plots horizontals at 200 mg toxic
!and 400 mg lethal doses.

SET WINDOW 0, 36, 0, 1100
LET tinitial = 0
LET tfinal = 36
```

```
LET t = tinitial
LET G = 1100
LET B = 0
LET numberofsteps = 3600
LET deltat = (tfinal-tinitial)/numberofsteps
FOR k = 1 TO numberofsteps
    LET Gprime = -0.1386*G
    LET Bprime = 0.1386*G - 0.1155*B
    LET deltaG = Gprime*deltat
    LET deltaB = Bprime*deltat
    PLOT t, G
    PLOT t, B
    PLOT t, 200
    PLOT t, 400
    LET t = t + deltat
    LET G = G + deltaG
    LET B = B + deltaB
NEXT k
END


!Program DOWHILE (True BASIC)

!Determines when amount of drug in bloodstream reaches 400 mg,
!or when amount of drug in bloodstream peaks.
!Plots both graphs up to point of interest.
!Uses Euler's method.

SET WINDOW 0, 20, 0, 1200
LET tinitial = 0
LET t = tinitial
LET G = 1100
LET B = 0
LET Gprime = -0.1386*G    !Compute G', B' for t = 0.
LET Bprime =  0.1386*G - 0.1155*B
LET deltat = 0.01
DO WHILE B < 400      !We'll stop when B = 400 (or when B > 400).
!DO WHILE Bprime > 0  !We'll stop when B' = 0 (or when B' < 0).
                     !This is the beginning of the DO-WHILE loop.
    LET deltaG = Gprime*deltat
    LET deltaB = Bprime*deltat
    LET t = t + deltat
    LET G = G + deltaG
    LET B = B + deltaB
    PLOT t, G                         !To make program run faster,
```

```
    PLOT t, B                       !hide these PLOT commands.
    LET Gprime = -0.1386*G
    LET Bprime = 0.1386*G - 0.1155*B
LOOP                      !This is the end of the DO-WHILE loop.  Just
                          !like NEXT k, it sends the computer back up
                          !to the top of the loop.
PRINT "t ="; t,"G ="; G,"B ="; B
END
```

About the Authors

Jodye Selco, professor of chemistry at the University of Redlands since 1987, earned her B.S. in chemistry from the University of California, Irvine, in 1979 and her Ph.D. in chemistry from Rice University in 1984. She is a physical chemist specializing in spectroscopy (meaning that she uses a lot more mathematics than other chemists!). She enjoys performing her Chemistry Magic Show at local schools, star-gazing, growing all her own vegetables, and making long commutes on the Southern California freeways.

Janet Beery, professor of mathematics at the University of Redlands since 1989, earned her B.S. in mathematics and English literature form the University of Puget Sound in 1983 and her Ph.D. in mathematics from Dartmouth College in 1989, specializing in permutation group theory. She has a new-found interest in history of mathematics and currently is writing historical modules with high school teachers. Since her undergraduate days, she has enjoyed reading novels without having to write papers about them, as well as traveling to rainy locales.

Jodye Selco and Janet Beery both enjoy teaching with technology, writing and assigning classroom projects, and helping students discover ideas for themselves.

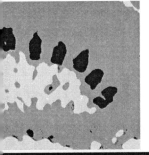

INTERDISCIPLINARY LIVELY APPLICATIONS PROJECT

AUTHORS:
Marie Vanisko
(Mathematics, Engineering,
Physics, and Computer
Science)
vanisko@carroll.edu

Carroll College
Helena, MT

Terry Johnson (Mathematics)
tjohnsog1@initco.net

Helena High School
Helena, MT

EDITORS:
David C. Arney,
Kathleen Snook, and
Steve Horton

CONTENTS:
Setting the Scene
Part 1: College Algebra
Part 2: Elementary
Statistics
Part 3: Introductory
Management Science
Instructor's Comments
and Solutions
About the Authors

Managing Health Insurance Premiums

MATHEMATICS CLASSIFICATIONS:
Part I: College Algebra
Part II: Elementary Statistics
Part III: Management Science, Operations Research

DISCIPLINARY CLASSIFICATIONS:
Health Information Management, Business, Ethics, Sociology

PREREQUISITE SKILLS:
Part I: Linear, logarithmic, and exponential functions,
          and curve fitting
Part II: Measures of central tendency and dispersion, confidence
          intervals, Chebyshev's inequality, curve fitting, and
          correlation coefficient
Part III: Measures of central tendency, curve fitting, and forecasting

PHYSICAL CONCEPTS EXAMINED:
Data on claims and premiums

MATERIALS INCLUDED:
Data in table form

COMPUTING REQUIREMENT:
Either a graphing calculator or a computer with spreadsheet,
computer algebra system, and/or statistical package. Part III is
most easily done with quantitative management software.

Contents
1. **Setting the Scene**
2. **Part 1: College Algebra**
3. **Part 2: Elementary Statistics**
4. **Part 3: Introductory Management Science**
Instructors' Comments and Solutions
About the Authors

1.   Setting the Scene

In the United States today, approximately 85% of the population has some type of public or private health insurance. In most cases, the coverage is through a group health plan. Each year, the group's premiums to maintain the existing plan usually have to be raised to cover anticipated expenditures for the following year. To limit the rate of increase in premiums, the group sometimes scales back coverage and/or explores ways to be more cost-effective.

The key advantage of group insurance is that all members in a group pay the same premium and have potentially the same benefits. Thus, no one member is penalized excessively for having higher-than-average medical costs.

Although most groups buy insurance coverage from insurance companies, a group that is large enough can set aside a reserve and become its own "miniature insurance company," which is to say that it becomes self-insured. Such a group frequently hires professionals to administer the plan, but the group retains more control over premiums than if they went through an insurance company. The data in this ILAP are taken from a self-insured group of approximately 1,000 individuals.

A natural concern of any group is how to absorb catastrophic expenses. Just as individuals can limit their liability by purchasing insurance that requires meeting deductibles before the insurance takes effect, groups and companies too can take out catastrophic insurance plans, called *re-insurance*, that do not take effect until a very large deductible is met. The group examined in this ILAP purchases re-insurance to cover catastrophic expenses on individuals (*specific coverage*) and also re-insurance to cover catastrophic expenses on the group as a whole (*aggregate coverage*). This strategy reduces the liability of the group and helps make the plan actuarially sound.

Assumptions

- The number of members in the group stays at 1,000.

- Each group member pays the same premium. (This is not completely realistic, because often a group charges different premiums for individual, family, and limited family coverages.)

Claims Record

Table 1 and **Figure 1** show data on claims for the group.

Table 1.
Historical information on dollar value of health insurance claims for the ILAP Group.

Year	Actual Claims
82–83	822,739
83–84	845,149
84–85	1,055,664
85–86	1,465,600
86–87	1,641,332
87–88	1,273,909
88–89	1,462,926
89–90	1,533,377
90–91	1,498,199
91–92	1,516,732
92–93	1,770,221
93–94	2,301,936
94–95	2,262,621
95–96	3,566,033
96–97	3,763,328

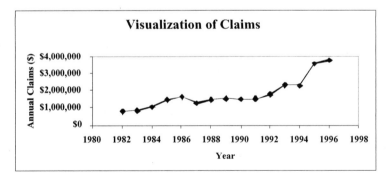

Figure 1. Graph of annual claims.

2. Part I: College Algebra Portion

Suppose that you are a member of this insurance group. You are quite dissatisfied with the fact that insurance premiums are rising, so you decide to examine the pattern of claims. Use the data in **Table 1** to complete the requirements below. All reports are to be submitted in written form with detailed explanations. Be prepared to discuss your solutions in class.

Requirement 1.

- Look at the data in **Table 1** and describe what you see reflected in the numbers.

- Review the assumptions made and determine if they are realistic.

- Why might the data be so volatile?

Requirement 2.

- Based on the actual claims and using the assumptions given, estimate the individual premium for each year so that the amount collected from premiums equals the amount of claims.

- Realizing that we can only guess at what the claims would be for the upcoming year, what might you suggest for possible premiums for individuals for the upcoming year and why? (Only an informal analysis is needed here.)

- Estimate the pattern in the scatter plot in **Figure 1** using two different models, one linear between the year and the total claim and one linear between the year and the natural log of the total claim.

- Which appears to be the better fit?

- Show that in the second model the relationship between the year (x) and the total claim (y) could be written in the form $y = ae^{bx}$.

- Show the graph of each model together with the scatter plot.

- For each model, estimate the total claims for the two years beyond the last dates given in **Table 1** and use these figures to suggest premiums.

Requirement 3.

- Compute the changes in claims from one year to the next and then compute the percentage changes. (%Change = Change / Original)

- From your percentage change data, compute each of the following measures of central tendency: arithmetic mean, median, and geometric mean.

- Which would you say gives the best measure of central tendency for these data? Justify your answer.

Requirement 4.

- Draw a scatterplot of percentage change versus time.

- Determine the line of best fit for this set of points and use it to predict what a member's monthly premium might be for the following year.

- Draw a scatterplot of the change in claims versus the previous year claims.

- Determine the line of best fit for this set of points and use it to predict what a member's monthly premium might be for the following year.

Requirement 5.

- If the change in claims is related in a linear fashion to the claims themselves, what is the relationship between the total claims and the time variable? Relate this back to what you investigated in **Requirement 2**.

- Discuss whether any of your measures of central tendency play a role in your models?

Requirement 6. Suppose that you work for the insurance group and are responsible for setting the participant premiums. Can you use the results of **Requirements 4** and **5** to help you? How?

Requirement 7. Imagine that you are the employer and you have a longstanding agreement to pay health insurance costs for your employees. Discuss what might you do to limit the cost of premiums under each of the following scenarios:

- You care more about profit than anything else, or

- you want to be fair to your employees.

Are these scenarios mutually exclusive? Be prepared to present your solutions to these requirements in class.

3. Part II: Statistics Portion

As an insurance group member who has had a statistics course, you decide to examine the data from a statistical perspective in an attempt to better quantify your predictions. Use the data in **Table 1** to complete the following requirements. All reports are to be submitted in written form with detailed explanations. Be prepared to discuss your solutions in class.

Requirement 1.

- Examine the assumptions and discuss their validity.

- How might you modify the problem to make it more realistic?

Requirement 2.

- Compute the changes in claims from one year to the next and then compute the percentage changes. (%Change = Change / Original)

- For the percentage change data, compute each of the following measures of central tendency: arithmetic mean, median, and geometric mean.

- Which would you say gives the best measure of central tendency for these data? Justify your answer.

- Represent the percentage change in claims using a box plot.

Requirement 3.

- Compute the standard deviation for the percentage change in claims.

- What assumption would you have to make to use these data to compute a 95% confidence interval for the percentage change in claims? Do these assumptions seem reasonable here? Why or why not?

- Assuming that these assumptions are valid, compute a 95% confidence interval for the percentage change in claims.

Requirement 4.

- Draw a scatterplot of percentage change versus time.

- Determine the least-squares line of best fit for this set of points and use it to predict the premium for the following year.

- Compute and interpret the coefficient of determination for this linear model. How would you make use of this coefficient of determination in describing the situation? What factors are not being taken into account?

Requirement 5.

- Use Chebyshev's inequality

$$P[|X - \mu| > k\sigma] \leq 1 - \frac{1}{k^2}$$

to give an upper bound on the likelihood that the percentage increase would exceed 50%.

- Why might Chebyshev's inequality be more appropriate here than the confidence interval that you generated in **Requirement 3**? Discuss the normality assumption or lack thereof.

Requirement 6.

- Using the upper bound determined in **Requirement 5** for the probability that the percentage increase exceeds 50%, translate that into the likelihood that the upcoming claim will exceed a certain amount.

- How might the re-insurance company use this bound in determining both the amount to charge the group for aggregate re-insurance and what the deductible would be?

- How could the insurance group make an attempt to lower their costs associated with re-insurance?

Requirement 7. Imagine that you are the employer and you have a longstanding agreement to pay health insurance costs for your employees. What might you do to limit the cost of premiums under each of the following scenarios:

- You care more about profit than anything else, or

- you want to be fair to your employees.

Are these scenarios mutually exclusive?

4. Part III: Management Science Portion

You have a background in management science and have been asked by the insurance group to analyze the pattern of their total claims, as well as their costs vs. premiums per individual over a twenty-four month period. You will use the data from **Table 1** and also the data from **Table 2**. All reports are to be submitted in written form with detailed explanations. Be prepared to discuss your solutions in class.

Requirement 1. Use each of the following time-series forecasting methods to determine what the claims will be for the upcoming year. Plot these models with a scatterplot of the data.

- 5-year moving average, experimenting with different weights;

- exponential smoothing;

- linear trend;

- linear trend with exponential smoothing.

Discuss which technique seems to be the most appropriate.

Requirement 2.

- Compute the changes in claims from one year to the next and then compute the percentage changes. (%Change = Change / Original)

- From the percentage change data, compute each of the following measures of central tendency: arithmetic mean, median, and geometric mean.

- Discuss why economists might suggest using the geometric mean. What type of model would this mean suggest?

Requirement 3. In reality, the number of participants in the group health plan changes slightly each month. **Table 2** gives a monthly breakdown of pertinent information over a 32-month period. As the group's advisor:

- Using line graphs, show the pattern of costs and the pattern of premiums. On a separate graph, show a scatter plot of the net gain (loss) to reserves.

- Discuss your recommendation for the amount of money that the group should have in their reserve fund.

- How will you handle the employer's not wanting to pay increased premiums?

- Address the employees' concern that their costs are rising and that a few individuals in the group are responsible for the huge increase they see coming.

Requirement 4. Imagine that you are the employer and you have a longstanding agreement to pay health insurance costs for your employees. What might you do to limit the cost of premiums under each of the following scenarios:

- You care more about profit than anything else, or

- you want to be fair to your employees. Are these scenarios mutually exclusive?

Requirement 5. The group's procedure for reimbursing participants for prescriptions was changed about five years ago. Before then, prescriptions were paid for by the participant, who could submit receipts to be reimbursed for part or all of the cost. Not all participants bothered to seek reimbursement. Under the new plan, the pharmacy is reimbursed directly for costs covered by the group plan, and the group member pays for only the non-eligible portion. This procedure serves the participants well (less paperwork!) but results in increased claims (and hence premiums), since now all prescription costs are automatically submitted. In fact, the impact on premiums was modest, but there was a side-effect of the change in policy:

- The group used information (from pharmacies' claims) about medications that many participants were taking to set up a wellness program, thus hoping to slow the growth of premiums in the long run.

- Although names were kept confidential, certain participants—particularly those taking antidepressants—felt that their privacy was invaded and that they were the target group for the wellness program.

Comment on the ethics of using information gathered in this way to set up a wellness program, and write a one- to two-page essay that addresses the issue of privacy vs. cost in this regard.

Table 2.
Average costs and average premiums over a 32-month period.

Year	Month	Participants	Avg. Cost	Avg. Premium	Net Total Difference
1995	6	984	212	239	26,568
	7	990	239	238	(990)
	8	990	243	238	(4,950)
	9	999	171	235	63,936
	10	1001	243	223	(20,020)
	11	1017	228	247	19,323
	12	1025	153	245	94,300
1996	1	1027	237	245	8,216
	2	1030	522	245	(285,310)
	3	1032	302	221	(83,592)
	4	1036	253	221	(33,152)
	5	1036	350	251	(102,564)
	6	1035	267	307	41,400
	7	1035	269	307	39,330
	8	1035	325	307	(18,630)
	9	1040	313	246	(69,680)
	10	1052	438	244	(204,088)
	11	1045	280	247	(34,485)
	12	1045	215	245	31,350
1997	1	1041	337	254	(86,403)
	2	1041	300	248	(54,132)
	3	1044	181	245	66,816
	4	1044	290	247	(44,892)
	5	1042	348	227	(126,082)
	6	1046	341	263	(81,588)
	7	1046	331	238	(97,278)
	8	1046	324	221	(107,738)
	9	1051	256	303	49,397
	10	1043	291	288	(3,129)
	11	1033	259	290	32,023
	12	1037	247	292	46,665
1998	1	1041	324	296	(29,148)
		Net loss to reserves for the period:			**(968,527)**

Title: **Managing Your Health Insurance Premiums**

Instructors' Comments and Solutions

Part I: College Algebra Portion

Requirement 2. For consistency, as well as ease in using particular software, you may wish to advise your students to code the years, for example, as 1 for 1982, ... , 15 for 1996. An Excel graph of the desired relationships is in **Figure S1**. The equations of the curves and the coefficients of determination regarding the linear relationships are also shown.

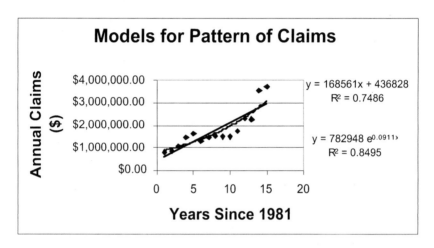

Figure S1. Models for claims.

The logarithmic model corresponding to the exponential equation would be approximately

$$\ln y = 13.5708 + .0911x.$$

Using these equations, the anticipated claims for the following two years would be as in **Table S1**.

Table S1.
Predicted claims.

Model	1997–1998 Prediction	1998–1999 Prediction
Linear	3,133,804	3,302,365
Exponential	3,363,261	3,684,044

Requirement 3. See **Table S2**.

Table S2.
Solution to **Requirement 3**.

Year	Actual Claims	Change	% Change	Geometric Mean Computations (%)
82–83	822,739			
83–84	845,149	22,410	2.7	102.7
84–85	1,055,664	210,515	24.9	124.9
85–86	1,465,600	409,936	38.8	138.8
86–87	1,641,332	175,732	12.0	112.0
87–88	1,273,909	−367,423	−22.4	77.6
88–89	1,462,926	189,017	14.8	114.8
89–90	1,533,377	70,451	4.8	104.8
90–91	1,498,199	−35,178	−2.3	97.7
91–92	1,516,732	18,533	1.2	101.2
92–93	1,770,221	253,489	16.7	116.7
93–94	2,301,936	531,715	30.0	130.0
94–95	2,262,621	−39,315	−1.7	98.3
95–96	3,566,033	1,303,412	57.6	157.6
96–97	3,763,328	197,295	5.5	105.5
		arithmetic mean	13.1	
		median	8.8	
		geometric mean	11.5	

Requirement 4.

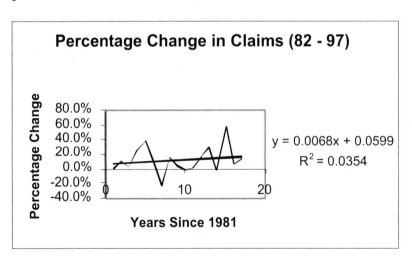

Figure S2. Year-to-year relative change in claims.

From the line of best fit to percentage change vs. time (**Figure S2**), we get

Expected claims = \$4,407,610, expected premium = \$367.30/member/month.

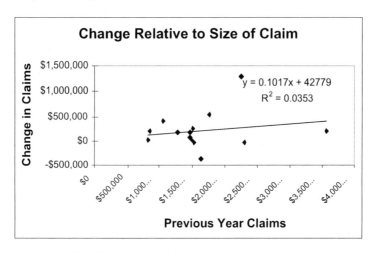

Figure S3. Change in claims vs. previous year's claims.

From the line of best fit to change in claims vs. previous year's claims (**Figure S3**), we get

Expected claims = \$4,188,837, expected premium = \$349.07/member/month.

Part II: Statistics Portion

Requirement 2. Refer to **Table S1** in **Part I** for the percentage changes and the measures of central tendency. Additional information needed for the boxplot includes the first quartile $Q_1 = 1.61\%$ and the third quartile $Q_3 = 22.86\%$. A boxplot of the percentage change data is shown in **Figure S4**.

Requirement 3. The standard deviation for the percentage change is 19.954%. If the assumptions of normality and random sampling were valid, a 95% confidence interval for percentage change would be approximately $13.1\% \pm 11.1\%$.

Requirement 4. Refer to **Figure S2** in **Requirement 5** for Part I.

Requirement 5. Since the Z-score for 50% is 1.85, Chebyshev's inequality would say that there is no more than a 29% chance of being that far away from the mean or farther. This statistic includes the possibility of landing on the other side of the mean as well.

Part III: Management Science Portion

Requirement 1. See **Table S3**.

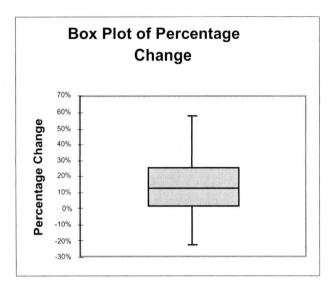

Figure S4. Boxplot of percentage change in claims.

Table S3.
Solution to **Requirement 1**.

Method	Prediction	R^2
5-year moving average	2,732,828	.52
Exponential smoothing	3,763,328	.73
($\alpha = 1$ works best)		
Linear trend	3,133,804	.75
Linear trend with exponential smoothing	4,230,655	.79
($\alpha = 0.75$, $\beta = 0.35$ works best)		

Requirement 2. Many economists would suggest using the geometric mean, resulting in the model

$$y = 822739(1.1147)^t,$$

where t represents the number of years after 1982.

Requirement 3. Figure S5 gives a chart showing the premiums (level curve) and the claims (jagged curve) per member each month.

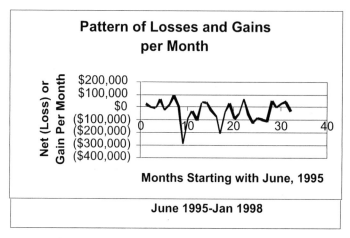

Figure S5. Premiums minus claims, on a monthly basis.

About the Authors

Marie Vanisko is a professor of mathematics at Carroll College in Helena, Montana, where she has taught for more than 30 years. As a co-director of the NSF Project InterMath at Carroll, she has been a primary mover in initiating the writing of interdisciplinary projects (ILAPs), and she has taken a lead role in instituting curricular reform in the undergraduate mathematics program. Marie is co-author of the technology supplement that accompanies the 10th edition of Thomas's *Calculus* (2001) and has served as a judge in COMAP's Mathematical Contests in Modeling at both the undergraduate and the high school levels. In Spring 2002, she will be a visiting professor at the Department of Mathematical Sciences at the U.S. Military Academy at West Point.

Gary Johnson has taught mathematics at the secondary level in Helena, Montana, for more than 30 years and has served as a financial planner and as a football and track coach as well. He has been active in curriculum innovation and calculus reform. Presently, he is directing the teaching of a college-level calculus/difference equations course at the high school level and also teaches selected mathematics courses at Carroll College.

INTERDISCIPLINARY LIVELY APPLICATIONS PROJECT

JTHORS:
 Yu
ansportation Studies)
_lx@tsu.edu

rrington Steward
athematics)

:as Southern University
uston, TX

ITOR:
vid C. Arney

Ramp Metering of Freeways

MATHEMATICS CLASSIFICATION:
 Algebra

DISCIPLINARY CLASSIFICATIONS:
 Traffic Engineering

PREREQUISITE SKILLS:
 Elementary algebra calculation
 Proportional rates and calculation
 Solving linear equation
 Elementary optimization

PHYSICAL CONCEPTS EXAMINED:
 Concept of freeway ramp control
 Independent ramp metering technique
 Integrated ramp metering technique

COMPUTER REQUIREMENT:
 Microsoft Excel or similar tool
 Ability to use calculator

Contents

Figure 1. Freeway traffic.

1.   Setting the Scene

Congestion in the national highways is growing far faster than the population (**Figure 1**). It's outpacing the growths in registered drivers and in vehicles on the road. It's growing faster even than the increasing number of miles Americans drive each year. The congestion problem has become so bad that if it is not controlled, it will considerably deteriorate the quality of life, since people are living in a society that so heavily relies on transportation. It will hinder economic development and disrupt the normal life of the entire society. Solving the congestion problem is one of the most urgent issues facing the nation's political leaders.

Congestion on a freeway usually induces more negative impact on an urban traffic network than congestion on a regular surface street. A freeway is defined as a divided highway with full control of access and two or more lanes for the exclusive use of traffic in each direction. Freeways provide uninterrupted flow with no signalized or stop-controlled at-grade intersections and with direct access prohibited to and from adjacent property.

One of the important parameters in operation of a freeway is its capacity, the maximum number of vehicles per hour that the freeway can accommodate. If traffic demand (the total number of vehicles per hour that attempt to enter a freeway segment) exceeds freeway capacity, the freeway becomes congested, since only the number of vehicles per hour equal to the capacity can actually pass through the freeway segment.

A technique widely used in big cities to reduce congestion on freeways is ramp control (**Figure 2**). Vehicles enter and leave a freeway through ramps, which are lengths of road providing an exclusive connection between a freeway and a surface street; vehicles enter via an entrance ramp (or on-ramp) and leave via an exit ramp (or off-ramp). Ramp control is the application of control devices such as traffic signals, signing, and gates to regulate the number of vehicles entering or leaving the freeway. The objective is to balance demand and capacity of the freeway so as to maintain smooth flow on the freeway.

Figure 2. Ramp control. Note the signals on either side of ramp, controlling access from ramp to freeway.

2. Independent Ramp Metering

The primary forms of freeway ramp control include entrance ramp metering and entrance ramp closure (an extreme case of ramp metering), used to limit the rate at which traffic can enter the freeway. The metering rate depends on the purpose of the metering. Normally, metering is used to eliminate congestion on the freeway when demand would exceed capacity, so the metering rate at a ramp would be based on the relationship between demand upstream of the ramp, capacity downstream of the ramp, and the volume of traffic desiring to enter the freeway at the ramp. Freeway capacity at the downstream of the ramp is determined by the capacity of that freeway section.

If the sum of upstream demand and ramp demand is less than or equal to downstream capacity, metering is not needed to prevent congestion. On the other hand, if the upstream freeway demand alone is greater than downstream capacity, metering at the ramp would not eliminate congestion; in this case, the entrance ramp closure should be applied. Otherwise, the metering rate is set equal to the difference between downstream capacity and upstream demand.

Example of Independent Entrance Ramp Metering

Figure 3 illustrates an independent ramp (no adjacent on- or off-ramps exist) where the upstream demand is 5,400 vehicles per hour (vph), the downstream capacity is 6,000 vph, and the ramp demand is 700 vph.

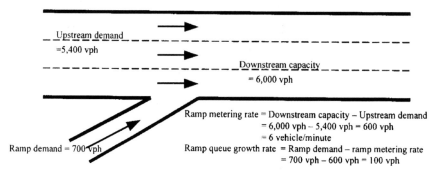

Figure 3. Example of independent entrance ramp metering rate calculation.

Since the total demand from the freeway upstream and ramp (5,400 + 700 = 6,100 vph) is greater than the downstream capacity (6,000 vph), ramp metering might be a feasible solution to prevent congestion. Therefore, if a metering rate equal to the difference between downstream capacity and upstream demand (6,000 − 5,400 = 600 vph, or 10 vehicles per minute) were used, the freeway could accommodate the upstream demand and maintain uncongested flow while also handling 600 vph of the ramp demand. Because the ramp demand

(700 vph) is 100 vph higher than the ramp metering rate (600 vph), a vehicle queue forms on the ramp at a rate of 100 vehicles per hour. For example, after 6 minutes of operation at this metering rate, there will be 10 vehicles waiting in the queue on the ramp.

3. Integrated Entrance Ramp Metering

Integrated ramp metering refers to expanding the independent entrance ramp metering to a series of entrance ramps. The metering rate for each of these ramps is determined in accordance with demand-capacity constraints at the other ramps as well as its own local demand-capacity constraint. Determining these metering rates requires the following information:

- mainline and entrance ramp demands,

- freeway capacities immediately downstream of each entrance ramp, and

- description of the traffic pattern within the freeway section to be controlled.

This information provides the basis for establishing the demand-capacity constraints of the entrance ramps and their interdependencies.

Given the required data, the fundamental procedure for computing metering rates involves five steps:

1. Start with the entrance ramp that is farthest upstream.

2. Determine the total demand (freeway demand immediately upstream of the ramp plus ramp demand) for the freeway section immediately downstream of the ramp.

3. Compare the total demand to the capacity of the downstream section and proceed as follows:

   (a) If the total demand is less than the capacity, metering is not required at this ramp by this demand-capacity constraint. Therefore, skip Step 4 and go immediately to Step 5.

   (b) If the total demand is greater than the capacity, metering is required at this ramp by the demand-capacity constraint. Therefore, proceed to Step 4.

4. Compare the freeway demand immediately upstream of the ramp to the capacity of the downstream section and proceed as follows:

   (a) If the freeway demand immediately upstream of the ramp is less than the downstream capacity, then the allowable entrance ramp volume (or metering rate) is set equal to the difference between the downstream capacity and the freeway demand immediately upstream of the ramp.

(b) If the freeway demand immediately upstream of the ramp is greater than or equal to the capacity, then the allowable entrance ramp volume is zero and the ramp must be closed. If the freeway demand immediately upstream of the ramp is greater than the capacity, the volumes permitted to enter at ramps upstream of the current ramp must be reduced accordingly. The total reduction in the allowable entrance ramp volumes upstream is equal to the difference between the freeway demand immediately upstream of the ramp and the downstream capacity, adjusted to account for that portion of the traffic entering upstream that exists before it reaches the downstream entrance ramp being closed.

5. Select the next entrance ramp downstream and go back to Step 2.

This procedure is illustrated by the following example.

Example of Integrated Ramp Metering

Figure 4 illustrates a freeway segment with three entrance ramps (and three off-ramps) for calculation of integrated ramp metering rates.

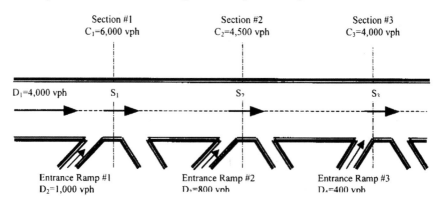

Figure 4. Illustration of integrated ramp metering for a freeway segment with three ramps.

Three consecutive entrance ramps are located along a length of a freeway section, which are named as Ramp 1, Ramp 2, and Ramp 3. Mainline freeway demand (D_1), three ramp demands (D_2, D_3, and D_4), and the capacities (C_1, C_2, and C_3) of three sections at the downstreams of three ramps are shown in the figure. The traffic pattern within the freeway section is described by the decimal fractions (A_{ij}) of vehicles that enter at various demand sources (Input i: $i = 1, 2, 3, 4$) and pass through three sections (Section j: $j = 1, 2, 3$), as shown in the **Table 1**. For example, $A_{12} = 0.9$ means that $9/10$ of the vehicles of the mainline freeway demand continue through Section 2 and $1/10$ of the vehicles exit on the first off-ramp.

Table 1.

A_{ij} values, decimal fractions of vehicles that enter at various demand sources and pass through the sections.

$i \setminus j$	Section 1	Section 2	Section 3
mainline freeway (Input 1)	1.0	0.9	0.8
Ramp 1 (Input 2)	1.0	0.6	0.4
Ramp 2 (Input 3)	—	1.0	1.0
Ramp 3 (Input 4)	—	—	1.0

Let X_i represent the allowable volume at input i (the metering rates for entrance ramps). Let S_j express the total demand at Section j. Since only the entrance ramp control is being considered and not freeway mainline control, the allowable mainline volume X_1 at Input 1 is set to equal to the mainline demand D_1. Further, the demand S_j at Section j can be computed from

$$S_j = \left(\sum_{i=1}^{j} A_{ij} X_i \right) + A_{j+1,j} D_{j+1}. \tag{1}$$

Determination of ramp metering rates for three ramps—the computation of the X_js—is shown in the following:

- Set $X_1 = D_1 = 4000$ vph.

- For Ramp 1 (consider Section 1):

    - Total demand $S_1 = A_{11}X_1 + A_{21}D_2 = (1.0)(4000) + (1.0)(1000) = 5000$ vph.
    - Compare the total demand to the capacity: $S_1 = 5000$ vph $< C_1 = 6000$ vph.
    - Therefore, metering is not required at Ramp 1. Thus, the allowable demand X_2 is the ramp demand and so $X_2 = 1000$ vph.

- For Ramp 2 (consider Section 2):

    - Total demand: $S_2 = A_{12}X_1 + A_{22}X_2 + A_{32}D_3 = (0.9)(4000) + (0.6)(1000) + (1.0)(800) = 5000$ vph.
    - Compare the total demand to the capacity: $S_2 = 5000$ vph $> C_2 = 4500$ vph.
    - Upstream mainline demand $= A_{12}X_1 + A_{22}X_2 = (0.9)(4000) + (0.6)(1000) = 4200$ vph.
    - Compare the upstream mainline demand to the capacity: Upstream mainline demand $= 4200$ vph $< C_2 = 4500$ vph.
    - Therefore, metering is feasible and the metering rate X_3 equals the difference between the capacity and the upstream mainline demand $= 4500 - 4200 = 300$ vph. This amount is also the allowable demand at Ramp 2. Therefore, $X_3 = 300$ vph.

- For Ramp 3 (consider Section 3):

  - Total demand:

$$S_3 = A_{13}X_1 + A_{23}X_2 + A_{33}X_3 + A_{43}D_4$$
$$= (0.8)(4000) + (0.4)(1000) + (1.0)(300) + (1.0)(400)$$
$$= 4300 \text{ vph.}$$

  - Compare the total demand to the capacity:

$$S_3 = 4300 \text{ vph} > C_3 = 4000 \text{ vph.}$$

  - Upstream mainline demand $= A_{13}X_1 + A_{23}X_2 + A_{33}X_3$
$$= (0.8)(4000) + (0.4)(1000) + (1.0)(300)$$
$$= 3900 \text{ vph.}$$

  - Compare the upstream mainline demand to the capacity:

Upstream mainline demand $= 3900 < C_3 = 4000$ vph.

  - Therefore, the metering rate for Ramp 3 is $4000 - 3900 = 100$ vph. As this is also the allowable demand at Ramp 3, we have $X_4 = 100$ vph.

- Conclusion:

  - Ramp 1: No control needed.
  - Ramp 2: Meter at 300 vph.
  - Ramp 3: Meter at 100 vph.

Linear Programming Formulation

The fundamental procedure for the integrated ramp metering can be formulated into a linear programming problem, which can be solved by using computer optimization software. For example, MS Excel has a function to solve the linear programming problem. The linear programming problem for the above example for integrated ramp metering can be formulated mathematically as follows:

- Objective function:
$$\max(4000 + X_2 + X_3 + X_4)$$

(maximize total freeway volume)

- Subject to:

  - downstream demand of Ramp 1 must be less than capacity at Section 1:

$$(1.0)(4000) + (1.0)(X_2) \le 6000;$$

– upstream demand of Ramp 2 must be less than capacity at Section 2:

$$(0.9)(4000) + (0.6)(X_2) \leq 4500;$$

– downstream demand of Ramp 2 must be less than capacity at Section 2:

$$(0.9)(4000) + (0.6)(X_2) + (1.0)(X_3) \leq 4500;$$

– upstream demand of Ramp 3 must be less than capacity at Section 3:

$$(0.8)(4000) + (0.4)(X_2) + (1.0)(X_3) \leq 4000;$$

– downstream demand of Ramp 3 must be less than capacity at Section 3:

$$(0.8)(4000) + (0.4)(X_2) + (1.0)(X_3) + (1.0)(X_4) \leq 4000;$$

– ramp metering rates must be positive and less than or equal to ramp demands:

$$0 \leq X_2 \leq 1000, \qquad 0 \leq X_3 \leq 800, \qquad 0 \leq X_4 \leq 400.$$

When this problem is solved using MS Excel, the results are found to be the same as from the manual calculation.

Requirement 1

A freeway segment has three lanes in one direction and a capacity of 2000 vph per lane. An independent entrance on-ramp is connected to this freeway. The upstream freeway demands and the on-ramp demands have been observed for three hours in the morning as shown in **Table 2**.

Table 2.
Freeway upstream demands and entrance ramp demands, in vph, for three morning hours.

	7 A.M.–8 A.M.	8 A.M.–9 A.M.	9 A.M.–10 A.M.
Freeway upstream	5660	6000	5100
Entrance ramp	720	880	700

Requirement 1a. Draw a sketch of this freeway–entrance ramp section.

Requirement 1b. What ramp control strategies (including no control, ramp metering, and ramp closure) should be used at the ramp for the three different hours in order to achieve the smoothest traffic flow on freeway?

Requirement 1c. If ramp metering is needed for **Requirement 1b**, what is the metering rate?

Requirement 1d. At 8:00 A.M., how many vehicles are waiting on the ramp?

Requirement 2

Figure 5 shows a freeway segment that has three lanes in one direction and a capacity of 1900 vph per lane. The freeway has a volume at the upstream of the first on-ramp of 5200 vph. It has been observed that 5% of the freeway upstream traffic exit at the first off-ramp ($A_{11} = 1.0$, $A_{12} = 0.95$).

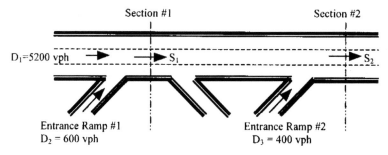

Figure 5. A freeway segment with 3 lanes, 2 on-ramps, and 1 off-ramp.

Using the same notation as earlier, please solve the following problems.

Requirement 2a. What is the total demand and capacity at Section 1?

Requirement 2b. What should be the metering rate at Ramp 1 to ensure that Section 1 on the freeway is not congested?

Requirement 2c. What is the total demand and capacity at Section 2?

Requirement 2d. What should be the metering rate at Ramp 2 to ensure that Section 2 on the freeway is not congested?

Requirement 2e. Formulate this ramp metering problem into a linear program and solve it using a software package (such as MS Excel).

Requirement 3

Figure 6 illustrates a freeway segment with a series of entrance and exit ramps. The freeway has three lanes with a directional capacity of 6000 vehicles per hour (vph), except for Section 3, where the capacity is reduced to 4000 vph due to construction work that blocks one lane. The freeway has a demand at the upstream of the first on-ramp of 3900 vph. Four entrance ramps have demands of 150, 2250, 200, and 700 vph respectively. The values of A_{ij}, the decimal fraction of vehicles entering at Input i that pass through Section j, are given by **Table 3**.

Using the same notation as earlier, please solve the following problems.

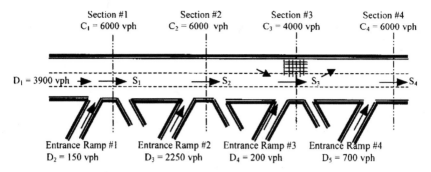

Figure 6. A series of ramps along a freeway segment.

Table 3.
Decimal fraction of vehicles entering at Input i that pass through Section j.

$i \setminus j$	1	2	3	4
1	1.00	0.95	0.90	0.80
2	1.00	1.00	0.85	0.80
3	—	1.00	0.90	0.85
4	—	—	1.00	0.95
5	—	—	—	1.00

Requirement 3a. Determine the appropriate control strategies on the four entrance ramps to achieve the smoothest flow on this freeway segment. If any downstream ramp must be closed, the upstream ramp metering rates must be readjusted so that the total demand at any section does not exceed its capacity.

Requirement 3b. Assume that the construction work at Section 3 is completed after one hour and the blocked lane is reopened. Determine the new control strategies on the 4 entrance ramps.

Requirement 3c. By the end of construction work at Section 3, what is the queue size on Ramp 3?

Requirement 3d. Formulate the question of **Requirement 3a** into a linear programming problem. Solve the problem using a software package (such as MS Excel).

Requirement 4

In a real-world urban traffic network, the capacity at a particular freeway section is usually a known variable upon the construction of the freeway. However, the traffic demands vary considerably over the locations and times. Problems in **Requirements 1–3** have been designed artificially so that the traffic demands from upstream of freeway and entrance ramps are given. In a real traffic network, however, the values of traffic demands have to be collected from the field in order to determine reasonable metering rates at the entrance ramp signals.

There are several methodologies to collect traffic demand data.

- The most widely used method is the Inductive Loop Detector (ILD). The ILDs are embedded under the surface of the pavement of highways. When a vehicle passes the location where an ILD is embedded, a traffic count is recorded. In this way, traffic demands can be found over a certain period of time.

- Sometimes manual counters are used. An individual is assigned to count the number of vehicles passing a point over a period of time.

- There are also other advanced methods, but they are beyond the scope of this project.

Figure 7 illustrates three entrance ramps and two exit ramps on a freeway (same as **Figure 4** except that the traffic demands are unknown). Assume that six traffic counters are placed at different sections of the freeway and the entrance ramps.

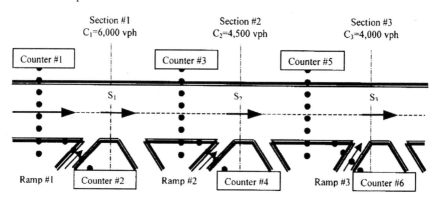

Figure 7. Illustration of integrated ramp metering for a freeway segment with three entrance ramps and six traffic counters.

Traffic counts at six counters in **Figure 7** are recorded every 5 minutes for a total of 30 minutes on a typical weekday at 6:00 P.M. to 6:30 P.M., as shown by **Table 4**.

With the above information, please answer the following questions:

Requirement 4a. Convert the collected 30-minute traffic counts at each counter to traffic demand in units of vehicles per hour.

Requirement 4b. Calculate the total number of vehicles that exit from off-ramp 1 during 6:00 to 6:30 P.M.

Table 4.

Traffic counts collected from six counters on the freeway and the entrance ramps.

Time period	Counter 1	Counter 2	Counter 3	Counter 4	Counter 5	Counter 6
6:00—6:05	317	79	351	30	308	25
6:05—6:10	322	81	346	32	312	29
6:10—6:15	336	85	359	29	326	31
6:15—6:20	341	84	364	32	335	33
6:20—6:25	355	87	374	41	343	36
6:25—6:30	333	85	364	38	328	34

Requirement 4c. Determine the optimal ramp metering rates at each ramp signal to maintain a smooth freeway operation.

Title: Ramp Metering for Freeways

Sample Solution

Requirement 1a. Please see **Figure S1**.

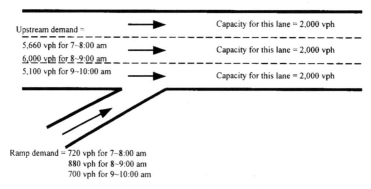

Figure S1. Solution to **Requirement 1a**.

Requirement 1b.

- For 7:00 A.M.–8:00 A.M.:

  Total demand = 5660 + 720 = 6380 vph > Capacity = (3)(2000) = 6000 vph.

  Since the upstream demand = 5660 vph < Capacity = 6000 vph, ramp metering is feasible and should be applied.

- For 8:00 A.M.–9:00 A.M.:

  Total demand = 6000 + 880 = 6880 vph > Capacity = 6000 vph.

  Since the upstream demand = 6000 vph = Capacity = 6000 vph, no vehicles should be allowed to enter the freeway from the ramp, and therefore the ramp should be closed.

- For 9:00 A.M.–10:00 A.M.:

  Total demand = 5100 + 700 = 5800 vph < Capacity = 6000 vph.

  Therefore, no control should be applied to the ramp.

Requirement 1c. For 7:00 A.M.–8:00 A.M., the ramp metering rate = difference between capacity and upstream demand = 6000 − 5660 = 340 vph.

Requirement 1d. For the time period of 7:00 A.M.—8:00 A.M., the ramp demand is 720 vph while the ramp metering rate is 340 vph. Therefore, the queue growth rate on the ramp = 720 − 340 = 380 vph. At 8:00 A.M., there will be 380 vehicles waiting on the ramp (spilling *far* back into the surface street!).

Requirement 2a. Total demand at Section 1 is

S_1 = upstream demand D_1 + Ramp 1 demand D_2 = 5200 + 600 = 5800 vph;

capacity at Section 1 is $C_1 = (3)(1900) = 5700$ vph.

Requirement 2b. Total demand at Section 1 is

$$S_1 = 5800 > 5700 = C_1 = \text{Capacity at Section 1.}$$

Since

$$\text{Upstream demand of Ramp 1} = (S_1 - D_2) = 5800 - 600 = 5200$$
$$< 5700 = C_1 = \text{Capacity at Section 1,}$$

ramp metering is feasible. Therefore, the ramp metering rate is $X_2 = 5700 - 5200 = 500$ vph.

Requirement 2c. Total demand at Section 2 is

S_2 = upstream demand D_1
  − upstream demand that exit from the first off-ramp$(.05)D_1$
  + Ramp 1 metering rateX_1 + Ramp 2 demandD_3
= 5200 − 260 + 500 + 400 = 5840 vph;

Capacity at Section 2 is $C_2 = (3)(1900) = 5700$ vph.

Requirement 2d. Total demand at Section 2 is

$$S_2 = 5840 > 5700 = C_2 = \text{Capacity at Section 2.}$$

Since

$$\text{Upstream demand of Ramp 1}(S_2 - D_3) = 5840 - 400 = 5440$$
$$< 5700 = C_2 = \text{Capacity,}$$

ramp metering is feasible. Therefore, the ramp metering rate is $X_3 = 5700 - 5440 = 260$ vph.

Requirement 2e. The linear programming formulation of the problem is as follows:

- Objective function: $\max(5200 + D_2 + D_3)$

- Subject to:
$$(1.0)(5200) + (1.0)(X_2) \leq 5700;$$
$$(0.95)(5200) + (1.0)(X_2) \leq 5700;$$
$$(0.95)(5200) + (1.0)(X_2) + (1.0)(X_3) \leq 5700;$$
$$0 \leq X_2 \leq 600, \qquad 0 \leq X_3 \leq 400.$$

- Using MS Excel to solve the problem results in $X_2 = 500$ and $X_3 = 260$.

Requirement 3a. The appropriate control strategies on the entrance ramps are determined as follows.

- Set $X_1 = D_1 = 3900$ vph.

- For Ramp 1 (consider Section 1):

  – Total demand $S_1 = A_{11}X_1 + A_{21}D_2 = (1.00)(3900) + (1.00)(150) = 4050$ vph.

  – Compare the total demand to the capacity $S_1 = 4050 < C_1 = 6000$ vph.

  – Therefore, metering is not required at this ramp, $X_2 = 150$ vph.

- For Ramp 2 (consider Section 2):

  – Total demand: $S_2 = A_{12}X_1 + A_{22}X_2 + A_{32}D_3 = (0.95)(3900) + (1.00)(150) + (1.00)(2250) = 6105$ vph.

  – Compare the total demand to the capacity: $S_2 = 6105$ vph $> C_2 = 6000$ vph.

  – Upstream mainline demand $= A_{12}X_1 + A_{22}X_2 = (0.95)(3900) + (1.00)(150) = 3855$ vph.

  – Compare the upstream mainline demand to the capacity: Since upstream mainline demand $= 3855 < C_2 = 6000$ vph, ramp metering is feasible.

  – Therefore, metering rate is $X_3 =$ difference between the capacity and the upstream mainline demand $= 6000 - 3855 = 2145$ vph.

- For Ramp 3 (consider Section 3):

  – Total demand: $S_3 = A_{13}X_1 + A_{23}X_2 + A_{33}X_3 + A_{43}D_4 = (0.90)(3900) + (0.85)(150) + (0.90)(2145) + (1.00)(200) = 5768$ vph.

  – Compare the total demand to the capacity: $S_3 = 5768$ vph $> C_3 = 4000$ vph.

  – Upstream mainline demand $= A_{13}X_1 + A_{23}X_2 + A_{33}X_3 = (0.90)(3900) + (0.85)(150) + (0.90)(2145) = 5568$ vph.

  – Compare the upstream mainline demand to the capacity: Since upstream mainline demand $= 5568 > C_3 = 4000$ vph, ramp metering is not feasible. The ramp must be closed and the upstream ramp metering rates should be readjusted.

  – The volume entering upstream of Ramp 3 must be reduced by $5568 - 4000 = 1568$ vph. Therefore, the metering rate at Ramp 2, X_3, must be reduced by $1568/A_{33} = 1568/0.90 = 1742$ vph. Thus, the metering rate at Ramp 2 is $X_3 = 2145 - 1742 = 403$ vph.

- For Ramp 4 (consider Section 4):

- Total demand:

$$S_4 = A_{14}X_1 + A_{24}X_2 + A_{34}X_3 + A_{44}X_4 + A_{54}D_5$$
$$= (0.80)(3900) + (0.80)(150) + (0.85)(403) + (0.95)(0) + (1.00)(700)$$
$$= 4282 \text{ vph.}$$

- Compare the total demand to the capacity: $S_4 = 4282 < C_4 = 6000$ vph.
- Therefore, metering is not required at this ramp, $X_5 = 700$ vph.

• Conclusion:

  - Ramp 1: No control needed.
  - Ramp 2: Meter at 403 vph.
  - Ramp 3: Closed.
  - Ramp 4: No control needed.

Requirement 3b. The appropriate control strategies on the entrance ramps are determined as follows.

• Set $X_1 = D_1 = 3900$ vph.

• For Ramp 1 (consider Section 1):

  - Total demand is

  $$S_1 = A_{11}X_1 + A_{21}D_2 = (1.00)(3900) + (1.00)(150) = 4050 \text{ vph.}$$

  - Compare the total demand to the capacity $S_1 = 4050 < C_1 = 6000$ vph.
  - Therefore, metering is not required at this ramp, $X_2 = 150$ vph.

• For Ramp 2 (consider Section 2):

  - Total demand is

  $$S_2 = A_{12}X_1 + A_{22}X_2 + A_{32}D_3$$
  $$= (0.95)(3900) + (1.00)(150) + (1.00)(2250)$$
  $$= 6105 \text{ vph.}$$

  - Compare the total demand to the capacity: $S_2 = 6105 > C_2 = 6000$ vph.
  - Upstream mainline demand $\quad = A_{12}X_1 + A_{22}X_2$
  $$= (0.95)(3900) + (1.00)(150)$$
  $$= 3855 \text{ vph.}$$

  - Compare the upstream mainline demand to the capacity: Since upstream mainline demand = 3855 < $C_2 = 6000$ vph, ramp metering is feasible.

– Therefore, metering rate

X_3 = difference between the capacity and the upstream mainline demand
$$= 6000 - 3855 = 2145 \text{ vph.}$$

• For Ramp 3 (consider Section 3): Capacity at Section 3 is increased to 6000 vph upon completion of the construction work

– Total demand:

$$S_3 = A_{13}X_1 + A_{23}X_2 + A_{33}X_3 + A_{43}D_4$$
$$= (0.90)(3900) + (0.85)(150) + (0.90)(2145) + (1.00)(200)$$
$$= 5768 \text{ vph.}$$

– Compare the total demand to the capacity: $S_3 = 5768 < C_3 = 6000$ vph.
– Therefore, metering is not required at this ramp, $X_4 = 200$ vph.

• For Ramp 4 (consider Section 4):

– Total demand:

$$S_4 = A_{14}X_1 + A_{24}X_2 + A_{34}X_3 + A_{44}X_4 + A_{54}D_5$$
$$= (0.80)(3900) + (0.80)(150) + (0.85)(2145) + (0.95)(200)$$
$$+ (1.00)(700)$$
$$= 5953 \text{ vph.}$$

– Compare the total demand to the capacity: $S_4 = 5953 < C_4 = 6000$ vph.
– Therefore, metering is not required at this ramp, $X_5 = 700$ vph.

• Conclusion:

– Ramp 1: No control needed.
– Ramp 2: Meter at 2145 vph.
– Ramp 3: No control needed.
– Ramp 4: No control needed.

Requirement 3c: Since Ramp 3 is closed and the demand at Ramp 3 is 200 vph during the construction work for one hour, by the end of construction at Section 3, the queue size on Ramp 3 will be 200 vehicles.

Requirement 3d. The linear programming formulation of the problem is as follows:

• Objective function: $\max(3900 + D_2 + D_3)$

- Subject to:

$$(1.00)(3900) + (1.00)(X_2) \leq 6000;$$
$$(0.95)(3900) + (1.00)(X_2) \leq 6000;$$
$$(0.95)(3900) + (1.00)(X_2) + (1.00)(X_3) \leq 6000;$$
$$(0.90)(3900) + (0.85)(X_2) + (0.90)(X_3) \leq 4000;$$
$$(0.90)(3900) + (0.85)(X_2) + (0.90)(X_3) + (1.00)(X_4) \leq 4000;$$
$$(0.80)(3900) + (0.80)(X_2) + (0.85)(X_3) + (0.95)(X_4) \leq 6000;$$
$$(0.80)(3900) + (0.80)(X_2) + (0.85)(X_3) + (0.95)(X_4) + (1.00)(X_5) \leq 6000;$$
$$0 \leq X_2 \leq 1500, \quad 0 \leq X_3 \leq 2250, \quad 0 \leq X_4 \leq 200, \quad 0 \leq X_5 \leq 700.$$

- Using MS Excel to solve the problem results in $X_2 = 150$ vph, $X_3 = 403$ vph, $X_4 = 0$ vph, and $X_5 = 700$ vph.

Requirement 4a.

Traffic at Counter 1	= 2004 vehicles/30 minutes	= 4008 vph
Traffic at Counter 2	= 501 vehicles/30 minutes	= 1002 vph
Traffic at Counter 3	= 2158 vehicles/30 minutes	= 4316 vph
Traffic at Counter 4	= 202 vehicles/30 minutes	= 404 vph
Traffic at Counter 5	= 1952 vehicles/30 minutes	= 3904 vph
Traffic at Counter 6	= 188 vehicles/30 minutes	= 376 vph

Requirement 4b.

$$\text{Off-Traffic at Exit Ramp 1} = \text{Traffic at Counter 1} + \text{Traffic at Counter 2}$$
$$- \text{Traffic at Counter 3}$$
$$= 4008 + 1002 - 4316$$
$$= 694 \text{ vph.}$$

Requirement 4c. The optimal metering rates at three entrance signals are determined as follows:

- Set $X_1 = D_1 = $ Traffic at Counter 1 = 4000 vph.

- For Ramp 1 (consider Section 1):

  – Total demand

  $$S_1 = \text{Traffic at Counter 1} + \text{Traffic at Counter 2} = 5010 \text{ vph.}$$

  – Compare the total demand to the capacity:

  $$S_1 = 5010 < C_1 = 6000 \text{ vph.}$$

  – Therefore metering is not required at Ramp 1. Thus, the allowable demand X_2 is the ramp demand and so $X_2 = $ Traffic at Counter 2 = 1000 vph.

- For Ramp 2 (consider Section 2):

  - Total demand

    S_2 = Traffic at Counter 3 + Traffic at Counter 4 = 4720 vph.

  - Compare the total demand to the capacity:

    $S_2 = 4720 > C_2 = 4500$ vph.

  - Upstream mainline demand = Traffic at Counter 3 = 4496 vph.
  - Compare the upstream mainline demand to the capacity:

    Upstream mainline demand = $4316 < C_2 = 4500$ vph.

  - Therefore, metering is feasible and metering rate is

    X_3 = difference between the capacity and the upstream mainline demand
    $= 4500 - 4316 = 184$ vph.

    This amount is also the allowable demand at Ramp 2. Therefore $X_3 = $ 184 vph.

- For Ramp 3 (consider Section 3):

  - Total demand:

    S_3 = Traffic at Counter 5 + Traffic at Counter 6 = 4280 vph.

  - Compare the total demand to the capacity:

    $S_3 = 4280 > C_3 = 4000$ vph.

  - Upstream mainline demand = Traffic at Counter 5 = 3904 vph.

  - Compare the upstream mainline demand to the capacity:

    Upstream mainline demand = $3904 < C_3 = 4000$ vph.

  - Therefore metering rate for Ramp 3 is $4000 - 3904 = 96$ vph. As this is also the allowable demand at Ramp 3, $X_4 = 96$ vph.

- Conclusion:

  - Ramp 1: No control needed.
  - Ramp 2: Meter at 184 vph.
  - Ramp 3: Meter at 96 vph.

Title: **Ramp Metering for Freeways**

Notes for the Instructor

This project is designed for students to practice basic algebra calculation as well as linear programming technique, using a real-world example for the freeway ramp metering techniques.

Freeway ramp metering is a very useful technique to manage the freeway traffic flows attempting to minimize the operational breakdown on freeways. This technique is becoming even more important today in light of the worldwide development and deployment of *Intelligent Transportation Systems (ITS)*. ITS is the application of new and emerging technologies such as computers, display, communications, process control, and sensors in transportation with objectives to reduce delays, travel time, fuel consumption, and vehicle emissions. Ramp metering is considered an integral component under the ITS umbrella.

Important to completing this project successfully is a clear understanding of the basic operations of ramps and ramp metering technique. While the project itself concerns only the determination of ramp metering rates, it would be useful if the instructor can provide some relevant information. The information that can be introduced includes the relation of ramp control/metering to other freeway management functions. Ramp control is closely related to other infrastructure elements in a freeway management system, as ITS movement has emphasized the benefits of integrated system elements. The following elements are directly or indirectly related to ramp control:

- **Surveillance—Vehicle Detection:** Detectors located in the freeway lanes generally have the purpose of providing input to incident detection algorithms and for system operation evaluation. Counters can also be used as input data in determining ramp metering rates. Counters located on entrance and exit ramps serve as input and output data in defining a closed system operation for estimating average delay in the system.

- **Surveillance—Closed-Circuit Television:** Closed-circuit television (CCTV) is used to detect and verify incidents in the overall surveillance subsystem. Cameras can also be used to fine-tune and monitor operation of individual metered ramps, precluding the necessity for on-site field observation.

- **Surveillance—Environmental Sensors:** Due to grades on ramps, it is often necessary to adjust ramp metering rates or terminate operation during extreme weather conditions such as icy or extremely wet roadway surfaces. Environmental sensors give early warning when such conditions exist.

- **HOV Treatments:** Preferential treatment of high-occupancy vehicles (HOV) at entrance ramps has been used successfully in several locations on entrance ramps. These systems have primarily involved a separate lane to bypass the ramp signal and a single-occupant-vehicle queue.

- **Information Dissemination:** Notification of travelers of ramp closures can be effected by either pre-trip information dissemination devices such as kiosks, Web site, and Community Access Television (CATV), or by on-road devices such as variable message signs or highway advisory radio. Other operational changes in ramp operations that may be of interest or assistance to travelers can also be communicated.

- **Communication:** Unless the controlled ramps are isolated and operate in a nonsystem mode, the communication subsystem must accommodate for the control, detection, and signal hardware.

- **Control Center:** While a ramp control system generally has the capability to operate in an isolated manner without supervision from a central or distributed master, most are interfaced to a central management system through communication system.

Students should be required to read **Setting the Scene** carefully and should clearly understand calculation procedure for ramp metering before they proceed to solving problems, which are designed to increase gradually in difficulty. Each requirement can be used as an independent problem, and all requirements together can be used as an integrated term project.

The requirements can have many variations. For example, the freeway upstream demand and ramp demands can be changed to different values; capacity at different sections can also be altered to reflect possible realistic conditions on freeways; and the number of on- and off-ramps can be added or reduced.

The requirements are designed to exercise only the basic ramp-metering calculation procedure. The rates resulting from the ramp-metering calculation in some cases may not be feasible or practical. In reality, the ramp-metering rates calculated for an integrated ramp-control system should be evaluated with respect to many practical considerations, such as:

- Metering rates of less than 180 to 240 vph (3 to 4 vehicles per minute) are not feasible, because drivers required to wait longer than 15 to 20 seconds at a ramp-metering signal often believe that the signal is not working correctly. They will therefore proceed against the red signal. Thus, if a metering rate of less than 180 to 240 vph is calculated, consideration should be given either to closing the ramp or to metering it at a higher rate.

- Practical maximum metering rates are about 900 vph for single-entry metering and approximately 1100 vph for platoon metering. Therefore, for a metering rate greater than the maximum for the metering type to be used, the setting should be less than or equal to the practical maximum rate, and the metering rates at the other entrance ramps should be adjusted accordingly.

- Metering rates equal to zero indicate that an entrance ramp closure is necessary. However, the closure of a particular entrance ramp may not be acceptable. Therefore, it may be necessary to increase a zero metering rate to some minimum acceptable rate.

- The procedure described for computing metering rates gives preference to traffic entering the system near the upstream end. Consequently, metering rates at entrance ramps downstream may be too restrictive to be acceptable to the motoring public. Therefore, it may be necessary to increase the metering rates computed for some of the downstream entrance ramps, and thus to reduce accordingly the metering rates for some of the upstream entrance ramps.

References

Carvell J.D., K. Balke, J. Ullman, K. Fitzpatrick, L. Nowlin, and C. Brehmer. 1997. *Freeway Management Handbook*. Report to Federal Highway Administration, FHWA–SA–97–064.

Transportation Research Board (TRB). 1998. *Highway Capacity Manual*. Special Report 209, 3rd Edition. Washington, DC: National Research Council.

Yu, L. 1995. Highway Traffic Operations. Course Notes for TMGT 862, Department of Transportation Studies, Texas Southern University, 1995.

About the Authors

Lei Yu is Associate Professor and Chair of the Transportation Studies Department at Texas Southern University. He is also a Changjiang Scholar of Northern Jiaotong University, China. He received a bachelor's degree in transportation engineering from Northern Jiaotong University, Beijing, China and a master's degree in production and systems engineering from Nagoya Institute of Technology, Nagoya, Japan. His Ph.D. degree in civil engineering was awarded by Queen's University in Kingston, Ontario, Canada.

As a professor at Texas Southern University, he has been teaching the courses in Highway Traffic Operations, Travel Demand Forecasting and Analysis, Transportation Design and Engineering, Computer Applications in Transportation, and Quantitative Analysis in Transportation. His research interests and expertise involve transportation network modeling, ITS related technologies and applications, dynamic traffic assignment and simulation, vehicle exhaust emission modeling, highway traffic control and operation strategies, travel demand forecasting models, and air quality issues in transportation.

In the past, Yu has served as the Principal Investigator (PI) of more than 20 research and consulting projects that were sponsored by various agencies. He has published numerous research papers in scientific journals and conference proceedings, plus project reports. Professionally, Dr. Yu is a licensed professional engineer (P.E.) in Texas and an active member of the Institute of Transportation Engineers (ITE), the American Society of Civil Engineers (ASCE), and the Transportation Research Board (TRB). He also is a member of numerous committees, councils, and task forces in regional, state, national, and international organizations.

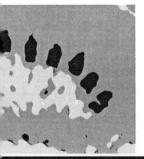

INTERDISCIPLINARY LIVELY APPLICATIONS PROJECT

THORS:
. Chidambaran (Finance)
ldi@tulane.edu

ι Liukkonen
thematics)
)math.tulane.edu

ine University
v Orleans, LA

TORS:
'id C. Arney
hleen Snook
ve Horton

The Lagniappe Fund:

The Story of Diversification and Asset Allocation

MATHEMATICS CLASSIFICATIONS:
Optimization, Probability

DISCIPLINARY CLASSIFICATIONS:
Economics and Finance

PREREQUISITE SKILLS:
Elementary probability and statistics (expected value, variance, and covariance, including their matrix formulations) and the method of agrange multipliers

PHYSICAL CONCEPTS EXAMINED:
Portfolio optimization, capital asset pricing model

MATERIALS INCLUDED:
Data at Website

COMPUTER REQUIREMENT:
Spreadsheet such as Excel or Lotus 123 strongly recommended.

Contents

1. Setting the Scene

You have been recently hired to manage the multi-million-dollar Lagniappe Fund, which is a pension fund for university professors.[1] Colleges that participate in the fund allow professors to make regular monthly contributions towards their retirement. The goal of the fund is to ensure that professors have an avenue to invest for retirement and manage these investments wisely. Each month, a small fraction of the investors in the fund retire and withdraw their money from the fund. It is the fund's policy to monitor the investment portfolio of their clients and offer investment advice.

A good investment policy is one that meets the needs of its investors. Consideration should be given to the number and mix of *assets*[2] to purchase and the number of years until the investor retires. The mix of assets that make up the fund's investments is called its *portfolio*. The Lagniappe Fund currently offers a choice between a bond portfolio and a stock portfolio. The two portfolios represent a strategy of investing in a broad variety of stocks and bonds, respectively, in amounts proportional to the total market values of the individual securities.

You have in hand a list of professors who have recently started their careers and are seeking advice on how to invest. You also have in hand a list of professors who will retire in two or three years, plus their current holdings in the stock and bond portfolios. How do you advise each group? Where

---

[1]lagniappe (lăn-yăp′, lăn′yăp′). 1. A small gift presented to a customer with his purchase by a store owner. 2. *Informal*. An extra or unexpected gift; a gratuity. —*American Heritage Dictionary of the English Language*, ed. by William Morris; Boston, MA: Houghton-Mifflin, 1976.

[2]Financial securities such as stocks and bonds are generically called *assets*. A financial asset allows you to invest money today in exchange for income in the future. Income over and above the amount invested is called the asset's return and is discussed in detail in this ILAP.

should they invest? In this ILAP, we develop the tools required to answer these questions scientifically.

2. The Business of Managing Money

Ever since the 1940 Investment Company Act established mutual funds, the way people save money has changed. Individuals have slowly moved money out of traditional investments, such as bank accounts and certificates of deposit (CDs), and have instead become investors in the stock and bond markets. A popular technique is to put money into stock and bond mutual funds that are managed by investment professionals. As a result, a large fraction of investors' wealth is entrusted to investment professionals, that is, professionally "managed." Investors entrust their money to mutual funds to reduce the costs of investing or if they believe that professionals can get better results than they could realize on their own.

A related phenomenon is the growth of the *pension fund* industry. Pension funds are similar to mutual funds in their operations, the difference being that the pension fund investors' objective is long-term and is geared towards having enough funds when they retire. This is increasingly a very important aspect of an individual's investment strategy. The Internal Revenue Service (IRS) encourages such activity by giving tax breaks to money that is invested in pension funds.

The goal of the pension funds is to invest the money wisely over a long period and ensure that the investment grows over time. The fund manager's objective is to get the best possible returns for the investor. This does not necessarily translate to getting the maximum returns possible. The fund manager also needs to factor in "risk," that is, the probability of potential losses in the financial markets. As we shall see later in this ILAP, it is easy to manipulate the risk and return levels of the investments. A fund manager should invest according to the wishes of the clients, and perhaps one of the most important tasks for the fund manager is to identify the needs of investors and manage the fund to meet their goals.

It is usual for a pension fund to manage more than one fund. These funds will differ in their risk levels; for example, a bond fund is thought to have low risk, whereas a stock fund is thought to have high risk. Based on their own risk preferences, investors either choose a fund in which to invest or choose a mix of the funds available. A good fund manager tracks the investment choices of the clients and is ready to provide advice on what is appropriate for them.

A fund manager also has to choose an investment strategy. A stock fund, for example, typically consists of a large number of stocks, and the manager has to decide which stocks to pick. A necessary first step is a proper analysis of each stock, to evaluate its potential return and its risk. Further, the fund manager has to decide how much to invest in the various stocks, relying on investment theory for guidance.

3.   Risk and Return: Economic, Mathematical, and Empirical Views

To make intelligent investment decisions, we must consider both the *return* and the *risk* of financial securities. Return and risk can be looked at from several viewpoints; each of these viewpoints has its place in investment theory.

In economic terms, the return on a financial security is defined as the amount of money earned over a fixed period of time, say one day or one month. The idea is to determine how much money an investor would earn if she were to buy the security today and sell it after a fixed time. This amount is called the security's *holding period return*.

The holding period return on a financial security, such as a share of FORD stock, held for N days, is

$$\text{Holding Period Return} = \frac{\text{Ending Price} - \text{Beginning Price} + \text{Dividends}}{\text{Beginning Price}}.$$

The holding period return is normally *annualized* by converting it to an yearly basis.[3] Doing so ensures that even with data over time frames of differing lengths, securities are compared on an even footing. The *annualized holding period return* on a financial security held for N days is[4]

$$\text{Annualized Holding Period Return} = \text{Holding Period Return} \times$$
$$\frac{\text{number of days in a year}}{N}.$$

Obviously, the return on the financial asset is not guaranteed at the time when it is purchased, since the price at which the investor can sell it is unknown. To deal with the uncertainty of returns, we bring probability theory to bear on the analysis of security prices and formulate the concepts of return and risk mathematically. We start by assuming that the return on a financial security i is a random variable R_i. We can then speak of the *expected return* $E[R_i]$, the variance σ_i^2, and standard deviation σ_i of the return on the security. The expected return is a basic measure of the level of the return, and the standard deviation is one of

---

[3]The annualized holding period return is actually equal to

$$(1 + \text{holding period return})^{(\text{number of days in a year})/N} - 1.$$

The equation given in the text is a commonly used first-order Taylor series approximation for small values of the holding period return. The annualized holding period return is our best guess of the annual return.

[4]For annualizing daily returns in the examples presented in this ILAP, the number of days in a year is taken to be the number of trading days in a year. The number of trading days is equal to the number of days in a year minus all weekend days and holidays for which the financial markets are closed. In 1996, for example, the number of trading days was 254. In many situations, however, the number of days in a year is set to the normal 365 days. Monthly returns are annualized assuming 12 months in a year.

the measures of risk used in finance. This standard deviation is often referred to as the security's *volatility*.

To estimate the expected return and variance of return of financial securities, we use historical data from the financial market. The assumption is that the probability distribution of returns will be the same in the future as it has been in the past and we can thus estimate the future probability distribution using historic data. We refer to this as the *empirical approach* to return and risk.

To make legitimate empirical calculations, we must set out a clear probability model. We first fix both a holding period (i.e., the time over which the security is held by the investor) and the time over which the holding period returns are calculated. For example, we might look at monthly returns for each month over the last two years. For each individual stock, we assume that the monthly returns, $R_1, \ldots, R_n$, are independent identically distributed random variables. We can calculate a series of monthly holding period returns for each stock traded in the market. We assume that for each stock the distribution of the monthly returns remains the same from month to month.

But the monthly returns of different stocks are in general correlated with one another, since the general economic conditions affecting one stock also affect other stocks. So we assume further that the joint distribution of all stocks under consideration also remains the same from month to month. In other words, if we collect the returns of the various securities for each month into a vector, those return data vectors can be used to estimate the individual and joint probability distribution of monthly returns for the immediate future. Our estimates of the expected return, volatility, covariance, and correlation, based on the historically observed returns, can be obtained in the standard way, as discussed in **Appendix A3**. Note that we can do similar calculations using data on daily returns or for some other holding period of our choice.

Table 1 gives data on the prices and dividends on the stock of FORD, EXXON, and IBM.

From the data, the return for FORD in January is

$$\frac{29.5 - 28.875 + 0.35}{28.875} = 0.0338$$

and the return in February is

$$\frac{31.25 - 29.5}{29.5} = 0.0593.$$

Note that the returns in January incorporate a dividend of $0.35 paid by the stock in January 1996.

Table 1.
Monthly price and dividend data for IBM, EXXON, and FORD.

DATE	IBM Prices	IBM Dividends	EXXON Prices	EXXON Dividends	FORD Prices	FORD Dividends
951229	91.375		81.125		28.875	
960131	108.5		80.25		29.5	0.35
960229	122.625	0.25	79.5	0.75	31.25	
960329	111.25		81.5		34.375	
960430	107.75		85		35.875	0.35
960531	106.75	0.35	84.75	0.79	36.5	
960628	99		86.875		32.375	
960731	107.5		82.25		32.375	0.385
960830	114.375	0.35	81.5	0.79	33.5	
960930	124.5		83.25		31.25	
961031	129		88.625		31.25	0.385
961129	159.375	0.35	94.375	0.79	32.75	
961231	151.5		98		32.25	

Table 2.
Monthly returns data for the S&P 500 Index and FORD.

DATE	S&P 500	FORD
960131	0.0326	0.0338
960229	0.0069	0.0593
960329	0.0079	0.1000
960430	0.0134	0.0538
960531	0.0229	0.0174
960628	0.0023	−0.1130
960731	−0.0457	0.0119
960830	0.0188	0.0347
960930	0.0542	−0.0672
961031	0.0261	0.0123
961129	0.0734	0.0480
961231	−0.0215	−0.0153
$E[R_{i,\text{Monthly}}]$	0.0159	0.0147
$\sigma_{i,\text{Monthly}}$	0.0313	0.0577
$E[R_{i,\text{Annualized}}]$	0.2344	0.1758
$\sigma_{i,\text{Annualized}}$	0.1083	0.1998

Table 2 shows the return for the Standard and Poors (S&P) 500 Index[5] and for FORD based on calculations similar to the one above.

The returns for EXXON and IBM are to be calculated as part of the requirements at the end of this section. The expected return and variance of FORD returns is calculated using equations in **Appendix A3**:

$$E[R_{\text{FORD}}] = \frac{1}{12} \sum_{t=1}^{12} R_{\text{FORD},t} = 0.0147,$$

$$\sigma_{\text{FORD}}^2 = \frac{1}{12-1} \sum_{t=1}^{12} (R_{\text{FORD},t} - E[R_{\text{FORD}}])^2$$

$$= \frac{1}{12-1} \sum_{t=1}^{12} (R_{\text{FORD},t} - 0.0147)^2 = 0.003328.$$

Standard deviation, the square root of variance, is therefore $\sigma_{\text{FORD}} = 0.05769$. The annualized values are $E[R_{\text{FORD,Annualized}}] = 0.1758$ and $\sigma_{\text{FORD,Annualized}} = 0.1998$.[6]

The relationship between the random returns (R_i, R_j) of two securities (i, j) is captured by the covariance C_{ij} and the correlation ρ_{ij} of the two random variables R_i and R_j. The covariance between FORD and the S&P 500 Index is calculated using the appropriate equation in **Appendix A3** and the data from **Table 2**:

$$C_{\text{FORD,S\&P}} = \frac{1}{12-1} \sum_{t=1}^{12} (R_{\text{FORD},t} - E[R_{\text{FORD}}])(R_{\text{S\&P},t} - E[R_{\text{S\&P}}])$$

$$= \frac{1}{12-1} \sum_{t=1}^{12} (R_{\text{FORD},t} - 0.0147)(R_{\text{S\&P},t} - 0.0159) = 0.000091.$$

The correlation coefficient is

$$\rho_{\text{FORD,S\&P}} = \frac{C_{\text{FORD,S\&P}}}{\sigma_{\text{FORD}}\sigma_{\text{S\&P}}} = \frac{0.000091}{0.05769 \times 0.03125} = 0.0504.$$

[5]The S&P 500 Index is a portfolio of 500 stocks that are traded on the New York Stock Exchange (NYSE), the American Stock Exchange (AMEX), or the NASDAQ Stock Exchange. The performance of the S&P 500 Index is thought to capture the performance of the financial markets and is closely related to the performance of the economy as a whole. As such, it is an important indicator of the health of the markets and the economy. The S&P 500 Index is, however, *not* a traded security by itself. The Index value at any time is the average cost of building the portfolio by trading each of the 500 stocks separately in amounts proportional to the total value of each stock in the market. Index funds, available through most major mutual fund houses, allow investors to effectively invest in the index by implementing an investment strategy involving a mix of stocks such that the returns on the index fund closely tracks the index itself.

[6]Annualized means and standard deviations represent the implied mean and standard deviation of the annual return. As a first-order approximation, the annual return R is equal to $r_1 + r_2 + \cdots + r_{12}$, where the r_is are independently identically distributed monthly returns. The annualized measures represent the uncertainty in the annual returns and are not simply rescaled values of the mean and standard deviation of monthly returns. We estimate the annualized mean and the annualized standard deviation by multiplying the monthly mean by 12 and the monthly standard deviation by $\sqrt{12}$, respectively.

Table 3 shows the correlation matrix for the S&P 500 Index, IBM, FORD, and EXXON. While the covariance matrix will be different whether we are using daily returns or annualized daily returns, the correlation matrix is the same in either case.

Table 3.
Correlation matrix for the S&P 500 Index, IBM, FORD, and EXXON.

	S&P 500	IBM	EXXON	FORD
S&P 500	1.0000	0.4993	0.5715	0.0504
IBM		1.0000	−0.0907	0.1926
EXXON			1.0000	0.0000
FORD				1.0000

Now suppose that we have a portfolio of n securities with returns $R_1, \ldots, R_n$ over a fixed time period, and that $w_1, \ldots, w_n$ are the fractions of our wealth—or weights—invested in the n securities. Then the return on the portfolio is simply the weighted average of the individual securities' expected returns, that is, $R_p = w_1 R_1 + \cdots + w_n R_n$. We assume that all the money available for investments is invested among the n securities, which is easily captured by the restriction that $w_1 + \cdots + w_n = 1$. It seems natural to assume also that each $w_i \geq 0$, but in fact we can allow $w_i < 0$ for some i. A negative holding in a security corresponds to a *short position*[7] in that security, where the investor sells the security at today's prices and promises to deliver the security at a later time when the investor can buy it at a different—hopefully lower—price. The expected return on the portfolio of n securities, $E[R_p]$, is the weighted average of the expected returns of the securities in the portfolio. Using the definitions in **Appendix A2**, it can be written as

$$E[R_p] = \sum_{i=1}^{n} w_i E[R_i].$$

The standard deviation σ_p of the portfolio is given by the equation governing the standard deviation of n random variables and depends on the correlation coefficient matrix for the n variables. The correlation matrix specifies the correlation coefficient between any two of the variables. Using the definition in **Appendix A2**, it can be written as:

$$\sigma[R_p] = \sqrt{\sum_{i=1}^{n} w_i^2 \sigma_i^2 + 2 \sum_{i=1}^{n} \sum_{j>i}^{n} \rho_{ij} w_i w_j \sigma_i \sigma_j}.$$

---

[7] A neat thing about financial securities is that you can sell what you do not have! Such a transaction is called a *short sale*. Essentially, investors can borrow the security from another individual with a promise to return it in due course and pay all the dividends on the security in the interim. The details are normally handled by the broker or trading firm, and the whole transaction is entirely transparent to the investor.

For example, when we have a portfolio of the S&P index and FORD stock that have returns $R_{S\&P}$ and R_{FORD}, respectively, we can find the expected return and standard deviation of the portfolio having weights $w_{S\&P}$ of the index and w_{FORD} of FORD. Knowing the expected return and standard deviation of $R_{S\&P}$ and R_{FORD} and the covariance $C(R_{FORD}, R_{S\&P})$, we can use how the expectation and the variance behave on linear combinations (see **Appendix A2**) to estimate $E[R_p]$ and σ_p. Consider an investment strategy that has 50% invested in the S&P index and 50% in FORD:

$$E[R_p] = w_{S\&P}E[R_{S\&P}] + w_{FORD}E[R_{FORD}]$$
$$= 0.5 \times 0.0159 + 0.5 \times 0.0147 = 0.0153,$$
$$\sigma_p^2 = w_{FORD}^2\sigma_{FORD}^2 + w_{S\&P}^2\sigma_{S\&P}^2 + 2\rho_{FORD,S\&P}w_{FORD}w_{S\&P}\sigma_{FORD}\sigma_{S\&P}$$
$$= 0.5^2(0.057692) + 0.5^2(0.031252)$$
$$+ 2(0.0504)(0.5)(0.5)(0.05769)(0.03125)$$
$$= 0.00111.$$

The standard deviation is therefore $\sigma_p = 0.0335$.

It is important to understand the limitations of what we are doing, so we close this section with comments on our basic assumptions and conventions.

• Our model assumes that a stock price at any time is the result of a previous base price plus subsequent random, independent, identically distributed daily price changes, or steps. This implies that stock prices follow a random walk, and our model is called the *random walk model*.

• Possible objections to the random walk model are:

  – The daily price changes may not be independent of one another, and
  – the distribution of the individual steps may change if the time period is too long.

  In fact, assuming independent daily price changes does provide a reasonable approximation to reality, but one needs to be careful about working with too long a time period or the distribution of daily returns may change significantly. In other words, the second objection is the more important liability for our model.

• We assume that the independent daily price changes (and so daily returns) of all stocks in our portfolio follow a fixed multivariate normal distribution. This assumption guarantees that the daily returns of any portfolio follow an ordinary univariate normal distribution.

• We ask you to *calculate* various parameters such as the expected value and standard deviation of various monthly or daily returns, and the correlations between them. "Calculate" is a misnomer, because you will be actually be *estimating* those fixed parameters—that is, you will arrive at estimates of the parameters and not at their true values. We justify this on two grounds:

- We want to keep the presentation simple, and

- it is common practice.

• Also to keep things simple, we ask you to deal with monthly, rather than daily, returns.

• When asked to calculate mean and standard deviation of monthly/daily returns, you should calculate the monthly/daily values and the annualized values, that is, you should also report the mean and standard deviation of the implied annual return. This is in keeping with the normal convention of reporting financial data on an annualized basis. Note that the annualized mean and standard deviation represent the uncertainty in the annual returns and are not simply rescaled values of the mean and standard deviation of monthly/daily returns (see Footnote 3).

• Finally, for the calculations, we strongly recommend that you use a spreadsheet or other matrix oriented computer package such as MATLAB or Mathematica. Many of the calculations, such as those involving correlation matrices, are very sensitive to random error. Therefore, although we generally report even intermediate answers to no more than 4 significant figures, it is best to keep the results of all calculations stored to many significant figures, and this is most easily accomplished by means of a package such as those mentioned above.

3.1   Requirements 1–3

Requirement 1. Consider the financial data on the stock of EXXON given in **Table 1**. The entries in the table represents the price of the stock at the end of each month from December 1995 to December 1996. The table also provides the dividend on the stock and the month in which the dividend is paid. Assume for the purpose of your calculation that the dividends are all paid on the last days of the months.

a. Calculate the returns on EXXON for each month in 1996. For each month, assume that the stock is purchased at the end of the previous month and sold at the end of the current month.

b. Calculate the expected return and standard deviation for EXXON.

c. Carry out similar calculations for FORD (i.e., duplicate the text calculations but keep the results to high precision). Consider an investment strategy in which 50% of your money is invested in each of FORD and EXXON. Calculate the return for the portfolio of the two stocks for each month in 1996. Using the returns series so generated, calculate the expected return and standard deviation on the portfolio.

Requirement 2. Consider the financial data on the S&P 500 Index given in **Table 2**. The entries in the table represents the return on $1 for each month in 1996 for the index. Use the results of **Requirement 1** for the returns on EXXON and carry out similar calculations for FORD.

a. Calculate the expected monthly return and the standard deviation of the monthly return on the S&P 500 Index.

b. Calculate the covariance and correlation matrix for the returns on the S&P 500 Index, FORD, and EXXON.

c. Consider an investment strategy in which 40% of the funds are invested in FORD and 60% is invested in EXXON. Calculate the mean and standard deviation of the monthly return for the portfolio using the correlation matrix and other results derived above.

d. Consider an investment strategy in which 60% of the funds are invested in the index (say through an Index fund) and 40% is invested in FORD. Calculate the mean and standard deviation of the monthly return for the portfolio using the correlation matrix derived above.

e. Assume that you invest equal dollar amounts among the three securities, that is, that the fraction of funds invested in each security is one-third. Calculate the expected monthly return and the standard deviation for this portfolio.

Requirement 3. This is an optional requirement for those familiar with Excel spreadsheets. Historical daily returns in 1996 for the S&P 500 Index and for the stocks of IBM, FORD, and EXXON are available at http://www.tulane.edu/ ~chiddi/ilaps.html in an Excel 97 worksheet. Using daily data, calculate the expected return and standard deviation for each of the securities and the correlation matrix for the three securities. Assume that you invest equal dollar amounts among the three securities, that is, that the fraction of funds invested in each security is one-third. Calculate the expected return and standard deviation for the portfolio in the following two ways:

• by using the portfolio equations, and

• by calculating the portfolio return each day and using the resulting returns vector.

4. The Strategy of Diversification: The Impact of Correlation and Covariance

In this section, we discuss the principles of diversification and its application as an investment strategy. *Diversification* is the strategy of dividing one's wealth among several securities and forming a portfolio, rather than investing it all in one security. Using the basic mathematical results from the previous section, we show that diversification allows us to achieve a lower standard deviation— a lower risk—compared to investing in individual securities. Diversification, therefore, is the key to extracting the maximum possible stock returns for a given level of risk.

A key insight that emerges is the following:

> *When the number of securities in the portfolio is very large, the standard deviation of the portfolio is almost completely determined by the covariances among the securities rather than by the standard deviations of the individual securities.*

Under such circumstances, a portfolio is said to be *completely diversified*. Thus, the part of the portfolio volatility arising from individual stocks can be nearly eliminated through diversification, whereas the part coming from the covariances between securities cannot be reduced.

4.1   Building Blocks of Portfolio Theory

To highlight the role of covariance in determining the standard deviation of the portfolio, we explore in detail the portfolio mathematics of a two-stock portfolio.

Consider a portfolio consisting of two securities, Stock 1 and Stock 2, with expected returns $E[R_1]$, $E[R_2]$, standard deviations σ_1, σ_2, and correlation ρ_{12} between the two returns. Without loss of generality, we assume that Asset 1 has higher standard deviation than Asset 2 (i.e., $\sigma_1 > \sigma_2$). Note that ρ_{12} is bounded, $-1 \leq \rho_{12} \leq 1$. If w_1 and w_2 are the fractional weights allocated to the two stocks, then the expected return and standard deviation of the portfolio are

$$E[R_p] = w_1 E[R_1] + w_2 E[R_2],$$

$$\sigma_p = \sqrt{w_1^2 \sigma_1^2 + w_2^2 \sigma_2^2 + 2\rho_{12} w_1 w_2 \sigma_1 \sigma_2}.$$

We make the restriction $w_1, w_2 \geq 0$ for the discussion in this section. In each of the three scenarios presented below, we calculate the weights on each of the stocks that minimize the standard deviation of the portfolio. We then discuss the implications for the minimum portfolio variance that can be achieved. A final section summarizes the results.

Diversification with Perfect Negative Correlation

$$\sigma_p = \sqrt{w_1^2 \sigma_1^2 + w_2^2 \sigma_2^2 + 2\rho_{12} w_1 w_2 \sigma_1 \sigma_2}$$

$$= \sqrt{w_1^2 \sigma_1^2 + w_2^2 \sigma_2^2 - 2 w_1 w_2 \sigma_1 \sigma_2}$$

$$= w_1\sigma_1 - w_2\sigma_2.$$

What does this tell us?

- σ_p is greatest when $w_1 = 1$.

- Portfolio variance is 0 (i.e., $\sigma_p = 0$) when $w_1\sigma_1 = w_2\sigma_2$. Since $w_1 + w_2 = 1$, the condition is that $w_1 = \sigma_2/(\sigma_1 + \sigma_2)$. The return corresponding to an asset or portfolio that has zero standard deviation is known with certainty and is therefore referred to as the *risk-free rate*.

- We can eliminate risk entirely from the portfolio and achieve the full benefits of diversification.

Figure 1a shows the trajectory of $(\sigma, E[R])$ corresponding to $\rho_{12} = -1$. The expected return and standard deviation of the two stocks are $E[R_1] = 0.22$, $E[R_2] = 0.15$, $\sigma_1 = 0.2$, and $\sigma_2 = 0.12$. The point where the two lines meet on the y-axis is the point representing the portfolio that has zero σ_p and a risk-free return of 0.17636.[8]

Diversification with Perfect Positive Correlation

$$\sigma_p = \sqrt{w_1^2\sigma_1^2 + w_2^2\sigma_2^2 + 2\rho_{12}w_1w_2\sigma_1\sigma_2}$$
$$= \sqrt{w_1^2\sigma_1^2 + w_2^2\sigma_2^2 + 2w_1w_2\sigma_1\sigma_2}$$
$$= w_1\sigma_1 + w_2\sigma_2.$$

What does this tell us?

- σ_p is greatest when $w_1 = 1$.

- σ_p is least when $w_1 = 0$.

- Portfolio standard deviation is simply a weighted average of the standard deviation of the individual securities and there are no benefits of diversification. If the objective is to minimize σ_p, we should simply invest in Asset 2.

Figure 1b shows the trajectory of $(\sigma, E[R])$ corresponding to $\rho_{12} = 1$. The expected return and standard deviation of the two stocks are $E[R_1] = 0.22$, $E[R_2] = 0.15$, $\sigma_1 = 0.2$, and $\sigma_2 = 0.12$. Note that the trajectory is linear between the two points representing the two securities.

---

[8]A risk-free return of 0.1763 or 17.63% is on the high side. Risk-free returns in the United States are traditionally in the range of 3%–6%. The calculations reflect what would happen if securities with the $E[R]$ and σ specified had a perfect negative correlation. As you would expect, such securities are not to be found in reality. The return on Treasury Bills sold by the U.S. government is a good example of risk-free securities.

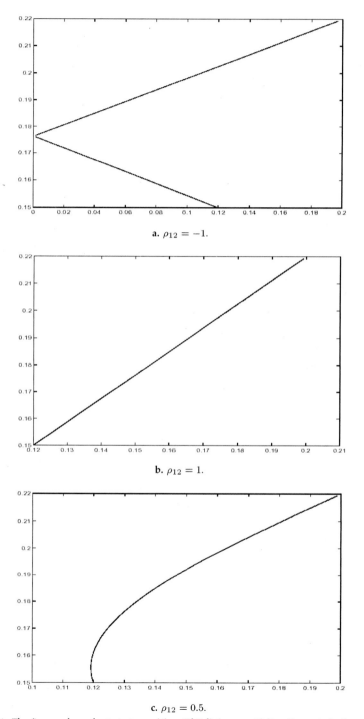

a. $\rho_{12} = -1$.

b. $\rho_{12} = 1$.

c. $\rho_{12} = 0.5$.

Figure 1. The figures show the trajectory of $(\sigma_p, E[R_p])$ for a portfolio of two stocks for various levels of ρ_{12}.

Diversification: $-1 \leq \rho_{12} < 1$

$$\sigma_p = \sqrt{w_1^2 \sigma_1^2 + w_2^2 \sigma_2^2 + 2\rho_{12} w_1 w_2 \sigma_1 \sigma_2}.$$

What does this tell us?

- σ_p is greatest when $w_1 = 1$.

- We can derive portfolio weights $w_1^{\min}, w_2^{\min}$ that minimize σ_p using the condition $w_1 + w_2 = 1$. A solution with $w_i > 0$ exists if $\rho_{12} < \sigma_1/\sigma_2$, where we assume $\sigma_2 < \sigma_1$. Such a portfolio is called the *minimum variance portfolio*, and

$$w_1^{\min} = \frac{\sigma_2^2 - \rho_{12}\sigma_1\sigma_2}{\sigma_1^2 + \sigma_2^2 - 2\rho_{12}\sigma_1\sigma_2}, \qquad w_2^{\min} = 1 - w_1^{\min}.$$

  The minimum variance portfolio has a variance lower than the lower of two variances (i.e., lower than σ_2^2) and a higher expected return than $E[R_2]$, since the portfolio return is the weighted average of $E[R_1]$ and $E[R_2]$.

- The minimum portfolio variance possible when $\rho_{12} > -1$ is strictly greater than 0. Therefore, only partial benefits of diversification can be achieved when $\rho_{12} > -1$, that is, risk can be reduced but not eliminated altogether.

  Figure 1c shows the trajectory of $(\sigma, E[R])$ corresponding to $\rho_{12} = 0.5$. The expected return and standard deviation of the two stocks are $E[R_1] = 0.22$, $E[R_2] = 0.15$, $\sigma_1 = 0.2$, and $\sigma_2 = 0.12$. Note that the left-most point on the trajectory represents the minimum variance portfolio with $E[R_p]^{\min} = 0.156$ and $\sigma_p^{\min} = 0.119$.

Summary

We conclude, therefore, that

- when $\rho = -1$, the portfolio manager can eliminate risk entirely and derive a risk-free rate;

- when $\rho = 1$, there are no benefits of diversification; and

- when $-1 < \rho < 1$, as is almost always the case, the portfolio manager can reduce the risk but not eliminate it altogether.

Example. Consider a portfolio of the S&P 500 Index and FORD stock. From **Table 2**, we have $E[R_{\text{FORD}}] = 0.0147$, $E[R_{\text{S\&P}}] = 0.0159$, $\sigma_{\text{FORD}} = 0.05769$, $\sigma_{\text{S\&P}} = 0.03125$, and $\rho_{\text{FORD,S\&P}} = 0.0504$. Since $\sigma_{\text{S\&P}}/\sigma_{\text{FORD}} > \rho_{\text{FORD,S\&P}}$, it is possible to find a portfolio with positive $(w_{\text{FORD}}, w_{\text{S\&P}})$ that minimizes σ_p. **Figure 2** shows the trajectory of $(\sigma, E[R])$. The values for $w_{\text{FORD}}^{\min}$ and $w_{\text{S\&P}}^{\min}$ are:

$$w_{\text{FORD}}^{\min} = \frac{\sigma_{\text{S\&P}}^2 - \rho_{\text{FORD,S\&P}}\sigma_{\text{FORD}}\sigma_{\text{S\&P}}}{\sigma_{\text{FORD}}^2 + \sigma_{\text{S\&P}}^2 - 2\rho_{\text{FORD,S\&P}}\sigma_{\text{FORD}}\sigma_{\text{S\&P}}}$$

$$= \frac{0.03125^2 - 0.0504 \times 0.03125 \times 0.05769}{0.03125^2 + 0.05769^2 - 2 \times 0.0504 \times 0.03125 \times 0.05769}$$

$$= 0.215$$

and $w_{S\&P}^{\min} = 1 - w_{FORD}^{\min} = 1 - 0.215 = 0.785$. Therefore, $E[R_p]^{\min} = 0.0157$ and $\sigma_p^{\min} = 0.028$, as indicated in **Figure 2**.

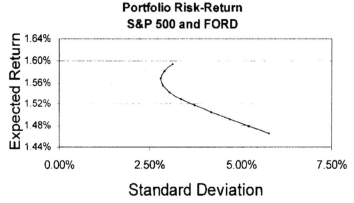

Figure 2. Portfolio risk-return. The figure shows the trajectory of $(\sigma_p, E[R_p])$ for a portfolio of the S&P 500 Index and FORD stock.

4.2   The Case of Multiple Securities

We can use linear algebra operations (i.e., matrix algebra) to calculate the expected return and standard deviation of return of a portfolio of securities. MATLAB and many other computer packages can carry out such operations quickly.

For example, consider the case of a portfolio of three securities. Let $E[R_1]$, $E[R_2]$, $E[R_3]$ be the expected returns; σ_1, σ_2, σ_3 the standard deviations of returns; and ρ_{12}, ρ_{13}, ρ_{23} the pairwise correlations. From the equations in **Appendix A2**, a portfolio of the three securities with weights (w_1, w_2, w_3)[9] has a portfolio standard deviation

$$\sigma_p = \sqrt{\sum_{i=1}^{3} w_i^2 \sigma_i^2 + \sum_{i}^{3} \sum_{j>i}^{3} 2\rho_{ij} w_i w_j \sigma_i \sigma_j}.$$

We can use this equation to calculate the portfolio weights that minimize σ_p for a given level of $E[R_p]$ subject to the conditions that $\sum_{i=1}^{3} w_i = 1$ and $E[R_p] = $

---

[9]We do not need to restrict w_i to be positive, as in **Section 4.1**. When w_i is negative, the portfolio has negative amounts of the security. This is possible in financial markets through short sales, as described in Footnote 4.

$\sum_{i=1}^{3} w_i E[Ri]$. However, it is easier and worth the effort to put this into vector-matrix terms. As a by-product, we obtain a formula for the minimum σ_p in the case of n securities. We use standard notation for matrices and (column) vectors; see **Appendix A1** for details.

For a portfolio of n securities, let $\vec{R}$ denote the $n \times 1$ column vector of returns on the n securities making up the portfolio. Let w denote the vector of weights. Then, letting $\mu = E[\vec{R}]$ be the column vector of expected returns and Σ be the covariance matrix of the returns on the n securities, we have for our particular portfolio the return $R_p = w^T \vec{R}$. It follows that $\mu_p = E[R_p] = w^T \mu$ and $\sigma_p^2 = w^T \Sigma w$. So in MATLAB, for example, we enter the weight vector w and estimates of μ and Σ, and there is little work left to estimate the expected return and the standard deviation of the portfolio returns.

We can also specify, in terms of each μ_p, what is the corresponding minimum possible σ_p. You will need this specification to work **Requirement 8** for this section. Suppose that we fix $\mu_p = E[R_p] = a$. What is the portfolio weight vector w that achieves minimal σ_p subject to this level of expected return? In linear algebra terms, we want to minimize $w^T \Sigma w$ subject to $w\mu = a$ and $w^T \vec{1} = 1$, where $\vec{1}$ is the column vector of ones. From **Appendix A4**, we have

$$w = c_1 \Sigma^{-1} \mu + c_2 \Sigma^{-1} \vec{1},$$

where c_1 and c_2 are obtained as follows. Let

$$A = \begin{bmatrix} \mu^T \Sigma^{-1} \mu & \mu^T \Sigma^{-1} \vec{1} \\ \mu^T \Sigma^{-1} \vec{1} & \vec{1}^T \Sigma^{-1} \vec{1} \end{bmatrix}.$$

Denote the column vector $[c_1, c_2]^T = A^{-1}[a, 1]^T$. Then $w = c_1 \Sigma^{-1} \mu + c_2 \Sigma^{-1} \vec{1}$. For this w, the minimum σ_p corresponding to $\mu_p = a$ is $\sigma_p = \sqrt{w^T \Sigma w}$.

4.3 Diversifiable and Nondiversifiable Risk

It is interesting to study the determinants of the portfolio variance, σ_P^2, as the number of securities in the portfolio increases. We can show that in a large portfolio only the covariances among the securities matter; the variances of individual securities do not contribute to portfolio variance. The individual security variances are said to be *diversified away* as the number of securities in the portfolio increases.

Suppose that we are dealing with k securities, the correlations of whose returns fall strictly between 0 and 1. How should we choose our weights? Let $R_1, \ldots, R_k$ be the returns of the k securities. Assume for simplicity that $E[R_i]$ and σ_i^2 are the same for all $i \in [1, k]$, and that ρ_{ij} and C_{ij} are also the same for each distinct pair i, j. Denote by $E[R]$ the common expected return, by σ the common standard deviation of return, by ρ the common correlation, and by γ the common covariance. Then,

$$E[R_p] = \sum_{i=1}^{k} w_i E[R] = E[R]$$

and

$$\sigma_p^2 = \sum_{i=1}^{k} w_i^2 \sigma_2 + \sum_{i=1}^{k} \sum_{j>i}^{k} 2\rho w_i w_j \sigma \sigma$$

$$= \sigma^2 \left[\sum_{i=1}^{k} w_i^2 + 2\rho \sum_{i=1}^{k} \sum_{j>i}^{k} w_i w_j \right].$$

No matter how you choose the weights w_i, the portfolio has a constant return $E[R]$. To minimize the risk of a portfolio that obtains this constant expected return, we need to minimize the expression in brackets subject to the condition $\sum_i w_i = 1$. Using Lagrange multipliers, we can show that the minimum is attained when $w_i = 1/n$ for each i, that is, we should invest an equal amount in each security (provided the simplifying assumptions that we made hold). Then

$$\sigma_p^2 = \frac{\sigma_2}{n} + \frac{n-1}{n}\gamma.$$

For large n, the first term goes to zero and $\sigma_p^2 \approx \gamma$. Thus, for large n, the common covariance γ, and not σ, is the dominant term in σ_p.

In general, the variances of all securities are not the same, nor are the pairwise correlations the same. However, the effect discussed above still holds; the correlations, or covariances, are the dominant terms in determining the variance of a portfolio.

To see the effect of increasing the number of securities graphically, imagine that the curve in **Figure 3** represents the expected return and standard deviation of return of combinations of two portfolios, not just a combination of two individual stocks. We can construct such curves for all two-pair combinations of securities and then use the portfolios so derived to draw yet another curve representing the best combinations of the portfolios, which are themselves best combinations of the individual securities, and so on. It can be shown that such an exercise results in a curve that looks like the trajectory in **Figure 3**. The upper-left portion of the boundary of the set of possible $(\sigma, E[R])$ is a concave downward curve and represents the portfolio combinations that minimize variance for a given level of expected return, using all the available securities. This curve is called the *efficient frontier*, since it gives the minimum variance possible for each particular level of expected return.

4.4   The Role of the Risk-Free Security

A special role is played by securities that have guaranteed returns and are considered to be *risk-free*. Treasury bills (T-bills), short-term obligations of the U.S. government, are a good example (and a proxy) of risk-free securities. We need to incorporate the role of risk-free securities into the portfolio investment decision.

E[R$_p$]

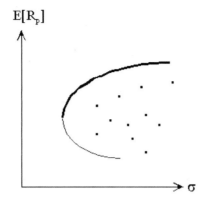

Figure 3. The efficient frontier for n risky assets. The figure shows the trajectory of $(\sigma_p, E[R_p])$ for a portfolio of n stocks. The upper portion of the curve represents the minimum possible variance for a specific level of expected return.

As in the previous section, assume that we are deciding how much money to invest in a portfolio of two securities: the risk-free security that has an expected return R_f and $\sigma_f = 0$, and a risky security T that has an expected return R_T and a prescribed σ_T. By definition, the covariance and correlation of the risk-free rate with the risky asset are zero, that is, $C(R_f, R_T) = \rho_{f,T} = 0$. Therefore, $\sigma_p = w_T\sigma_T$. Since $E[R_p] = w_f R_f + w_T E[R_T]$ and $w_f + w_T = 1$, we have

$$E[R_p] = R_f + \frac{\sigma_p}{\sigma_T}\left(E[R_T] - R_f\right).$$

Figure 4 shows that the above equation represents a straight line through points representing the risk-free security and the risky security T in $(\sigma, E[R_p])$ coordinates. All combinations of the two securities with positive weights lie in between the two points. To the right of the point representing the risky security T, the line represents portfolio combinations in which the weight on the risk-free asset is negative but the weight on the risky security is positive.

Treasury bills are bonds, and investments in them require the investor to pay money now and receive the money and added interest in the future. This makes the transaction tantamount to a loan—in this case, a loan to the government. On the other hand, a short sale of T-bills represents a cash inflow at the time of the sale with a promise to repay the money at a fixed time in the future, with interest added; this is equivalent to borrowing money. Because of the special nature of risk-free securities, therefore, investing in them is referred to as *lending* and short-sales are referred to as *borrowing*. Borrowing in order to invest is called *leverage*, and investors to the right of T are said to be investing *on the margin*. For all portfolio combinations to the right of T, we have $\sigma_p > \sigma_T$; leverage transactions are very risky. In practice, there are limits to how much you are allowed to borrow to invest (for example, you can have no more than 50% of portfolio value as borrowings); this limit is referred to as a *margin limit*.

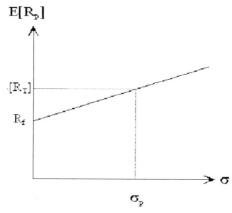

Figure 4. The role of the risk-free rate. The figure shows the line segment from the riskless security R_f and the risky security T in $(\sigma_p, E[R_p])$ coordinates.

Example. Assume that the risk-free security currently pays an interest of 0.005% per month (or an annual percentage return of 6%). How should you invest in a portfolio of FORD and the risk-free security such that you will be able to earn a monthly return of 2%? We know from the data in **Table 2** that the monthly expected return on FORD is $E_{\text{FORD}} = 0.0147$ and $\sigma_{\text{FORD}} = 0.0577$. Using the portfolio return equations, we get

$$0.02 = w_f \times 0.005 + w_{\text{FORD}} \times 0.0147.$$

Since $w_f = 1 - w_{\text{FORD}}$, we have $w_{\text{FORD}} = 1.546$ and $w_f = -0.546$. That is, we should borrow an additional 54.6% of the amount of money available for investments and invest the entire sum in FORD. We can see that

$$\sigma_p = 1.546 \times \sigma_{\text{FORD}} = 1.546 \times 0.0577 = 0.0892.$$

4.5   Requirements 4–8

Requirement 4. From the data on the stocks of FORD and EXXON calculated in **Requirement 1**, determine the minimum risk portfolio comprised of these two stocks. Assume that positive amounts of money have to be invested in each of the two stocks.

Requirement 5. For FORD and EXXON, calculate $(\sigma_p, E[R_p])$ as w_1 varies from 0 to 1 (keep in mind that $w_2 = 1 - w_1$). Sketch the trajectory of $(\sigma_p, E[R_p])$ as w_1 varies from 0 to 1. Use the data calculated in **Requirement 1**.

Requirement 6. Assume that a risk-free security is available that has a return of 0.005 per month (or an annual percentage return of 6%). Draw a graph that shows portfolio combinations of the risk-free security and EXXON. If the aim is to invest in a portfolio that has a return of 2.5% per month, what weight do you choose for each security? If the aim is to invest in a portfolio that has a standard deviation of 0.025, what weight do you choose for each security?

Requirement 7. Prove that the weights on the minimum variance portfolio are such that

$$w_1^{\min} = \frac{\sigma_2^2 - \rho_{12}\sigma_1\sigma_2}{\sigma_1^2 + \sigma_2^2 - 2\rho_{12}\sigma_1\sigma_2}$$

and $w_2 = 1 - w_1$. Verify the condition $\rho_{12} < \sigma_2/\sigma_1$, where $\sigma_1 > \sigma_2$, required for both $w_i > 0$.

Requirement 8. For a portfolio of the S&P Index, FORD, and EXXON:

a. Plot the set of $(\sigma, E[R])$ arising from all possible portfolios of our three instruments.

b. Identify the efficient frontier of this set of portfolios.

c. Verify your observations above by calculating the minimum variance portfolio for any particular level of return, using the formula for minimum σ_p given at the end of **Section 4.2** (middle, p. 103). If possible, produce a plot of minimum σ_p vs. $E[R_p]$ in the σ–$E[R]$ plane.

5. Mean-Variance Efficiency and Asset Allocation

In this section, we discuss the selection of an optimal investment portfolio. We assume that the desirability of any investment is characterized by its expected return and its risk, $E[R]$ and σ. An optimal portfolio obtains the minimum level of risk for the level of return.

We attack this problem geometrically and plot the set of all possible points $(\sigma, E[R])$ in the σ–$E[R]$ plane. By convention, the horizontal axis is the σ axis and the vertical axis is the $E[R]$ axis. We first consider certain geometric properties of this set and then argue that under mild plausible conditions an optimal point (and hence at least one optimal portfolio) exists. Finally, we develop a practical way to find this point.

5.1 Risk-Aversion and the Mean-Variance Framework

It is plausible to assume that investors prefer a higher expected return and would prefer not to have risk. For example, when we have to choose between getting $100 versus getting $1,000, we of course prefer the $1,000. And we prefer to get $1,000 with certainty rather than have a 25% chance of getting the $1,000 depending on whether it will rain tomorrow. The problem arises when we are given the choice of getting $100 with certainty or getting $1,000 with a 25% chance (which would you choose?). We rely on *utility theory*, the economics of personal preferences, to guide us.

We model the utility that investors receive from a combination of $\sigma, E[R]$ by using an equation such as

$$U(R) = E[R] - \frac{1}{2}A\sigma^2,$$

where A is called the *degree of an investor's risk-aversion* and takes on a value such as 2 or 3. The equation implies that:

- If two portfolios have the same risk (i.e., the same σ), then the portfolio with the higher expected return (i.e., the larger $E[R]$) is preferred.

- If two portfolios have the same expected return $E[R]$, then the one with smaller risk σ is preferred.

- Investors trade off expected return with standard deviation in choosing be- tween securities when one has both a higher standard deviation and a higher expected return.

Investors differ in the trade-offs they make on risk and return. An investor with high A is said to be highly *risk-averse* and would require a large amount of expected return for even small increases in σ. On the other hand, investors with low levels of risk-aversion (i.e., low A) can be induced to switch into risky investments by offering them a smaller increase in the expected return.

Why is this relevant to our problem? Well, the degree of risk-aversion decides the kind of portfolio that an investor will choose. Investors who are younger and have many years left in their careers are considered to be less risk-averse, since they can handle a potential loss better. Such investors could choose to invest in portfolios that have higher expected return even if they have higher risk. On the other hand, investors who are closer to retirement cannot handle losses very well—they are, after all, depending on their investments for the money that they need in retirement. Such investors are probably not interested in risky investments even with a promise of a higher expected return.

In our subsequent discussion, we talk of investors preferring higher returns and lower risk, and of investors having low levels of risk-aversion versus high levels of risk-aversion, without explicitly mentioning the above model. The model provides the necessary background for the discussion.

5.2   Asset Allocation

For each point (σ_p, μ_p) in the set of possible portfolios on the efficient frontier shown in **Figure 3**, we can add the entire line segment from R_f to (σ_p, μ_p). **Figure 5** shows the line segment for three such portfolios, T, A, and B.

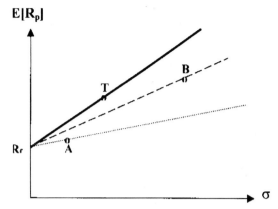

Figure 5. Optimal portfolio choice. The figure shows the line segment from the riskless security R_f and risky securities T, A, and B, in $(\sigma_p, E[R_p])$ coordinates.

For any portfolio combination on the lines from R_f to A (or to B), it is possible to find a portfolio on the line from R_f to T with a higher expected return or a lower standard deviation. Investors can therefore maximize their utility by choosing combinations of R_f and T rather than combinations with A or B, in spite of the fact that A has a lower standard deviation than T and B has a higher expected return than T. In picking an appropriate portfolio from the efficient portfolios, it is important to maximize the slope of the line that links R_f to the risky portfolio, which is $(\mu_p - R_f)/\sigma_p$. The slope is maximized at the point T where the line segment is tangent to the efficient portfolio, as show in **Figure 6**. Portfolios that lie on this line represent the best combinations of μ_p and σ_p that investors can achieve given the set of risky assets and the riskless asset.

Except for the point T, the tangency portfolio, other portfolios of risky assets with minimum variance for a given expected return are not good candidates for investment. Some combination of the risk-free security and the tangency portfolio gives higher expected return for the same level of variance.

We refer to the line through R_f and T as the *capital allocation line*. Investors cannot be better off (in utility terms) than picking combinations of the risky portfolio T and the risk-free security. The amounts that investors choose to invest in the risk security and the risk-free security depend entirely on the level of standard deviation, or risk, that is deemed acceptable. The investment decision therefore boils down to the amount of investments that are "allocated" to the risk-free asset, with the remaining amount invested in the risky asset T. There is no doubt whatsoever which risk-asset investors should choose. The portfolio T represents some optimal combination of the risky securities in the market and can be determined using linear algebra: Maximize $(\mu_p - R_f)/\sigma_p$ subject to $\sum_{i=1}^{n} w_i = 1$. Assuming, as in practice, that this optimal portfolio weight vector w occurs at a critical point of the objective function subject to the

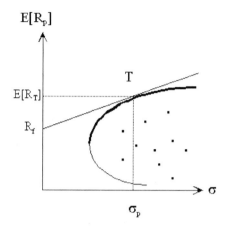

Figure 6. The efficient frontier and the line segment through R_f. The figure shows the line segment from the riskless security R_f and the tangential portfolio T in $(\sigma_p, E[R_p])$ coordinates.

side condition, then from **Appendix A4** we see that

$$w = \frac{\Sigma^{-1}\left(\mu - R_f\vec{1}\right)}{\left(\mu - R_f\vec{1}\right)^{T}\Sigma^{-1}\vec{1}}.$$

This formula will be needed to work **Requirements 11** and **12** at the end of **Section 5.3**.

5.3   Market Equilibrium

Our discussion so far has not involved more than one investor's decision-making. When we aggregate the behavior of all investors and study the effect of their interaction, we have to evaluate equilibrium in the financial markets. We start first with a set of assumptions:

- Investors have homogenous expectations—all investors agree on the expected return, variance, and covariance of the securities in the market.

- Investors are risk averse but are allowed to have varying levels of risk-aversion.

- Investors are atomistic—they trade individually, and their individual trades do not affect the prices of assets in the market.

Under these assumptions, we can prove that all investors will agree on the efficient frontier and the capital allocation line discussed in the previous sections. All investors also agree on what constitutes the tangency portfolio T.

All investors therefore allocate their investments exactly the same between the two assets, the risk-free rate and the risky asset T. We can show under these conditions that the tangency portfolio is made up of all the available risky securities and that the amount of money invested in each asset is proportional to the total value of that security (number of outstanding shares of security $i \times$ price per share) in the market, referred to as its *market capitalization*.

Since all agree on the constitution of the portfolio and it contains all the securities in the market, the tangency portfolio in equilibrium is called the *market portfolio*, denoted M. The market portfolio has an expected return $E[R_m]$ and standard deviation σ_m. When the tangency portfolio is explicitly set to be the market portfolio, the capital allocation line is renamed the *capital market line* (CML). Our theory then implies that all investors will allocate money between the risk-free rate and the market portfolio and that the amount they choose depends on their aversion to risk. The optimal investment strategy is therefore to pick a portfolio that satisfies the equation

$$E[R_p] = R_f + \frac{\sigma_p}{\sigma_T} \left(E[R_m] - R_f \right).$$

An interesting implication of the model is that the expected return on each security in the market is a linear function of the expected return on the market:

$$E[R_i] = R_f + \beta_i \left(E[R_m] - R_f \right).$$

Since the equation determines the equilibrium expected return for securities, it is called the *security market line (SML)*. The parameter

$$\beta_i = \frac{C(R_i, R_m)}{\sigma_m^2}$$

measures the sensitivity of the securities returns to the return on the market. The expected return on securities is determined by using either historical data or an estimate of the security beta and an estimate of the return on the market. The CML, SML, and beta make up one of the fundamental tools in finance for the valuation of financial assets. The model derived above is referred to as the *capital asset pricing model (CAPM)*.

A direct implication of the security market line equation is that the uncertainty of returns on the market portfolio is the single most important uncertainty in determining the expected return of securities. Note that σ_i does not appear in the equation for β and that only the covariance term seems to be important. These facts are due directly to the portfolio diversification result that we derived in the previous sections, where we showed that only the covariance terms were important in determining the σ of a portfolio when the number of securities in the portfolio is large. This gives rise to alternative definitions of risk, and we can write

Total Risk of the security = Market Risk + Firm-Specific Risk, where the firm's total risk is captured by σ and the market risk of the firm is captured by β. In a large portfolio, firm-specific risk is said to be *diversified away* and is not important in determining the expected risk of the security. Therefore, the expected return on the asset is determined by those factors that affect the market and is not affected by firm-specific shocks. Things that affect the market are "systematic"—they affect the market as a whole.

5.4   Requirements 9–12

Requirement 9. Suppose that in addition to stocks, we also invest in a Treasury bill that has an expected return (referred to as its *yield*) of 6% and a standard deviation of zero. What are the expected value and standard deviation of the return on our portfolio invested 35% in FORD, 55% in EXXON, and 10% in Treasury bills?

Requirement 10. Assume that the rate of riskless lending and borrowing is $R_f = .005$. On your plot in **Requirement 5** of **Section 4.5**, draw the line tangent to the set of possible $(\sigma, E[R])$ passing through the point $(0, R_f)$. Identify the point $(\sigma, E[R])$ corresponding to the tangency portfolio of FORD and EXXON.

Requirement 11. Apply the formula given at the end of **Section 5.2** (p. 110) to find the weights w_i corresponding to the tangency portfolio of FORD and EXXON. Verify that for this portfolio, $(\sigma, E[R])$ is the same as the point that you obtained graphically in the preceding **Requirement 10**.

Requirement 12. Consider your plot in **Requirement 8** of **Section 4.5** of all possible (σ_p, μ_p) for portfolios of the S&P 500 Index, FORD, and EXXON. Draw the tangent line to this set that passes through the point $(0, .005)$ and estimate from your plot the point (σ, μ) corresponding to the optimal portfolio of the S&P 500 Index, FORD, and EXXON. Now use the formula at the end of **Section 5.2** (p. 110) to determine the weights w_i corresponding to this optimal portfolio. Verify that for these weights, (σ_p, μ_p) is the same point of tangency that you obtained graphically.

6.   Lessons for the Portfolio Manager

Our discussions in this ILAP imply that the investment decision boils down to a decision on how much money should be invested in the risk-free security and how much in the market portfolio. Therefore:

- It is important is to determine the level of risk that the investor wants and can tolerate.

- The rules of thumb used by professional investment advisors are:

- If the level of risk that the investor can bear is lower—for example, if she is close to retirement—then the amount that she should invest in stocks is lower.

- If the level of risk that the investor can bear is higher—for example, if she is 21 and has many years of future expected wages—then she can invest a greater percentage of her securities in stocks.

- The alternatives that the Lagniappe Fund offers, a bond portfolio that behaves like the risk-free security and a stock portfolio that behaves like the market as a whole (e.g., a portfolio that tracks the S&P 500 Index), are optimal. The Lagniappe Fund should also allow investors to invest in any proportion that they choose between the two funds.

- The fund manager is deemed to be successful if the stock fund is able to get the same level of return as the market portfolio for the same level of market risk. In theory, it is possible to get a slightly higher return than the market portfolio if the fund manager is able to pick the better-performing stocks and avoid the bad performing stocks. Note, however, that the higher return should be at the same level of risk and not at a higher risk level.

6.1   Requirements 13–15

You have in your hand the list of professors who will retire in five years and their portfolio holdings, that is, the amounts of money that they have invested in Lagniappe's Bond Portfolio and the amount of money they have invested in Lagniappe's Stock Portfolio. You also have a list of the professors who are just starting their careers. Assume that you would like to act in the best interest of these professors. Based on what you learned in this ILAP:

Requirement 13. What would you look for in the holdings of the professors who will retire, in order to catch potential problems before they happen? Write a brief letter to a professor who you think has not invested wisely.

Requirement 14. What would your advice be to a professor who is just starting out on her career? Write a brief letter to her giving your recommendation on how she should invest.

Requirement 15. This is an optional requirement for those familiar with the Internet and willing to do some open-ended research. There is no one right answer to this requirement. Visit the Web pages of a large retirement pension fund company (such as Fidelity, Vanguard, or TIAA-CREF) and check what advice they give to their clients. Act the roles of a newly-hired professor and that of a professor who is about to retire and find the companies' recommendations. Write a report based on your findings.

Web addresses:
  Fidelity Investments: `http://www400.fidelity.com:80/`
  Vanguard: `http.://www.vanguard.com/`
  TIAA-CREF: `http://www.tiaa-cref.org/`

References

Bodie, Z., A. Kane, and A.J. Marcus. 1999. *Investments*. 4th ed. New York: Irwin-McGraw Hill.

Chamberlain, G. 1983. A characterization of the distributions that imply mean-variance utility functions. *Journal of Economic Theory* 29: 185–201.

Chong, E.K.P., and S.H. Zak. 1996. *An Introduction to Optimization*. New York: Wiley.

Elton, E., and M. Gruber. 1997. *Portfolio Theory*. 3rd ed. Englewood Cliffs, NJ: Prentice Hall.

Huang, C. F., and R.H. Litzenberger. 1988. *Foundations of Financial Economics*. Englewood Cliffs, NJ: Prentice Hall.

Merton, R. 1972. An analytical derivation of the efficient portfolio frontier. *Journal of Financial and Quantitative Analysis* 7: 1851–1872.

Roll, R. 1977. A critique of the asset pricing theory's tests. *Journal of Financial Economics* 4: 129–176.

Sharpe, W. 1964. Capital asset prices: A theory of capital market equilibrium under conditions of risk. *Journal of Finance* 19: 425–442.

Appendix: Mathematical Background

We collect here concepts from probability, statistics, and linear algebra that are used in this ILAP. We assume the reader is familiar with most or all of these. Our intent is not to provide an introduction to probability and statistics but rather to standardize concepts and notation. In **Appendix A4**, we provide derivations of two optimization formulas we use, for the benefit of those readers familiar with linear algebra and Lagrange multipliers.

A1. Matrices and Vectors

This section is intended to fix notation that is used in the ILAP. An $m \times n$ matrix A is a rectangular array of real numbers with m rows and n columns, denoted

$$A = \{a_{ij}\}_{i=1,\dots,m;j=1,\dots,n}.$$

An n-vector v is thought of as a point in R^n and also as an $n \times 1$ column matrix. This enables us to apply a matrix A to a vector v simply by taking the matrix product Av.

The *matrix product* AB of an $m \times n$ matrix A and an $n \times p$ matrix B is the $m \times p$ matrix C whose ijth coordinate is given by

$$c_{ij} = \sum_{k=1}^{n} a_{ik} b_{kj}$$

for each $i = 1, \dots, m$ and $j = 1, \dots, n$.

The *transpose* of the $m \times n$ matrix A is the $n \times m$ matrix A^T whose ijth coordinate is

$$a_{ij}^T = a_{ji}$$

for each $i = 1, \dots, m$ and $j = 1, \dots, n$. In other words, the rows of A are the columns of A^T. The transpose operation obeys the equation

$$(AB)^T = B^T A^T.$$

The $n \times n$ *identity matrix* I has entries 1 on its diagonal and entries 0 elsewhere. It satisfies

$$AI = A \qquad \text{and} \qquad IB = B$$

for all $m \times n$ matrices A and all $n \times p$ matrices B.

Finally, for a square $n \times n$ matrix A, the *inverse matrix* is the $n \times n$ matrix C such that

$$AC = I = CA,$$

if such a matrix C exists. Such a matrix C exists when various conditions are met, such as when the determinant of A is nonzero. When such a C exists, it is unique; we write $C = A^{-1}$. We have the properties

$$(AB)^{-1} = B^{-1} A^{-1}.$$

A2. Concepts from Probability

The *probability model* of any chance experiment includes a sample space S, with some or all of the subsets of S designated as events, and a probability rule or measure P that assigns to each event its probability. A *random variable* X is a variable whose value depends on the outcome of the experiment, so it is simply a function on the events of the sample space.

For random variables X that arise in practice, we have either a mass function f_X (for discrete X) or a density f_X (for continuous X). For the discrete case, we have $P(x \in A) = \sum_{x \in A} f_X(x)$ for every event A and the *expected value* $E[X] = \sum x f_X(x)$. Similar formulas hold for $E[X^2]$, $E[(X - c)^2]$, etc. For the continuous case, $P(X \in A) = \int_A f_X(x)\,dx$ for every event A and the expected value $E[X] = \int x f_X(x)\,dx$, while again similar formulas hold for $E[X^2]$.

No matter how expected value is defined, it can be used to introduce additional important concepts. If we denote the expected value or mean of X by $E[R_X]$, then the *variance* of X is defined as $V(X) = E[(X - E[R_X])^2]$, and the *standard deviation* of X is $\sigma_X = \sqrt{V(X)}$.

In practice, for a function ϕ of two variables and a pair X, Y of random variables, we can define the expected value $E[\phi(X, Y)]$ as $\sum \phi(x, y) f_{X,Y}(x, y)$ or $\int \phi(x, y) f_{X,Y}(x, y)\, dx dy$, where $f_{X,Y}$ is the joint mass function or joint density, as appropriate. Then the *covariance* is $C(X, Y) = E[(X - E[R_X])(Y - E[R_Y z])]$ and the *correlation* is $\rho(X, Y) = C(X, Y)/\sigma_X \sigma_Y$.

Intuitively, the mean $E[R_X] = E[X]$ is the center of mass of the distribution of X, while $V(X)$ and σ_X measure the spread of the distribution of X. Finally, the correlation measures the strength of linear association between X and Y.

Assume that X, Y, Z are arbitrary random variables and a, b, c are scalars. Also assume that all given expectations exist and are finite. Some specific properties that we need are:

$$E(c) = c,$$
$$E(aX + bY) = aE(X) + bE(Y),$$
$$V(aX + b) = a^2 V(X),$$
$$C(X, Y) = \rho(X, Y)\sigma_X \sigma_Y,$$
$$C(X, Y) = C(Y, X),$$
$$C(aX + bY, Z) = aC(X, Z) + bC(Y, Z),$$
$$V(aX + bY) = a^2 V(X) + 2ab C(X, Y) + b^2 V(Y).$$

In the special case that X and Y are independent, we have

$$E(XY) = E(X)E(Y),$$
$$V(X + Y) = V(X) + V(Y),$$
$$C(X, Y) = 0.$$

For the case of n random variables $X_1, X_2, \ldots, X_n$, with means $\mu_1, \mu_2, \ldots, \mu_n$, standard deviations $\sigma_1, \sigma_2, \ldots, \sigma_n$, covariances $C(X_i, X_j) = \sigma_{ij}$, and weights $w_1, w_2, , \ldots, w_n$, the formulas can be written as

$$E\left(\sum_{i=1}^{n} w_i X_i\right) = \sum_{i=1}^{n} w_i \mu_i,$$

$$V\left(\sum_{i=1}^{n} w_i X_i\right) = \sum_{i=1}^{n} w_i^2 \sigma_i^2 + 2\sum i = 1^n \sum_{j>i}^{n} w_i w_j \sigma_{ij}.$$

Recall that $\sigma_{ij} = \rho_{ij}\sigma_i \sigma_j$, where ρ_{ij} is the correlation. The formula for variance can also be written as

$$V\left(\sum_{i=1}^{n} w_i X_i\right) = \sum_{i=1}^{n} w_i^2 \sigma_i^2 + 2\sum i = 1^n \sum_{j>i}^{n} \rho_{ij} w_i w_j \sigma_i \sigma_j.$$

In vector notation, let $\vec{X} = [X_1, X_2, \ldots, X_n]^T$ be the vector of n random variables, $\vec{\mu} = [\mu_1, \mu_2, \ldots, \mu_n]^T$ be its mean vector, and σ be its covariance matrix (whose diagonal terms are σ_i^2 and the ijth term is $\sigma_{ij} = C(X_i, X_j)$). Let the vector $\vec{w} = [w_1, w_2, \ldots, w_n]^T$ be a vector of weights, where w_i is the weight on the random variable X_i. The linear combination $w_1 X_1 + \cdots + w_n X_n$ can be written as $\vec{w}^T \vec{X}$, and the above formulas can be written more compactly as

$$E\left(\vec{w}^T \vec{X}\right) = \vec{w}^T \vec{\mu}, \qquad V\left(\vec{w}^T \vec{X}\right) = \vec{w}^T \sigma \vec{w}.$$

A3. Estimating Parameters

In this ILAP and in financial practice, we have available samples, or several independent observations, from the random variables that we are studying. For example, if X and Y are stock prices, we may have n daily or monthly closing prices $x_1, \ldots, x_n$ and $y_1, \ldots, y_n$ from those stocks. We may then want to estimate the means $E[R_X]$, $E[R_Y]$, the standard deviations σ_X, σ_Y, the covariance $C(X, Y)$, and the correlation $\rho(X, Y)$.

To estimate $E[R_X]$ and σ_X, we use the standard estimates

$$\bar{x} = \frac{\sum_i x_i}{n},$$

$$s_x = \left(\frac{\sum_i (s_i - \bar{x})^2}{n - 1}\right)^{1/2}.$$

To estimate the covariance and correlation, we use

$$\hat{C}(X, Y) = \frac{\sum_i (x_i - \bar{x})(y_i - \bar{y})}{n - 1},$$

$$\hat{\rho}(X, Y) = \frac{\hat{C}(X, Y)}{s_x s_y}.$$

A4. Applications of Lagrange Multipliers

Under mild differentiability conditions, a local extremum x_0 of the objective function $f(x)$ subject to the side conditions $g_1(x) = 0, \ldots, g_k(x) = 0$ satisfies those side conditions and

$$\nabla f(x_0) = c_1 \nabla g_1(x_0) + \cdots + c_k \nabla g_k(x_0)$$

for some unknown scalars $c_1, \ldots, c_k$ (the *Lagrange multipliers*). For details, see an advanced calculus book or Chong and Zak [1996, Chapter 19].

We apply this technique of Lagrange multipliers to obtain two formulas used in **Section 4.2** and **Section 5.2** of this ILAP.

In Section 4.2

We want to find the vector w minimizing $w^T \Sigma w$ subject to $w^T \mu = a$ and $w^T \vec{1} = 1$, where a is a prescribed real number and $\vec{1}$ is a vector of ones. Here we assume that Σ is strictly positive definite. Clearly, $w^T \Sigma w \to \infty$ as $\|w\| \to \infty$, so we need check only critical points via the method of Lagrange multipliers.

The gradient of the objective function here is $2\Sigma w$ and the gradients of the side conditions are μ and $\vec{1}$. Therefore, at the critical point, we have $\Sigma w = c_1 \mu + c_2 \vec{1}$ for some scalars c_1, c_2 to be determined. So

$$w = c_1 \Sigma^{-1} \mu + c_2 \Sigma^{-1} \vec{1}.$$

Taking the side conditions into account, we get

$$a = c_1 \mu^T \Sigma^{-1} \mu + c_2 \mu^T \Sigma^{-1} \vec{1}$$
$$\vec{1} = c_1 \vec{1}^T \Sigma^{-1} \mu + c_2 \vec{1}^T \Sigma^{-1} \vec{1}.$$

Solving these equations for c_1, c_2 gives the desired result.

In Section 5.2

We want to maximize $(\mu_p - R_f)/\sigma_p$ subject to $w^T \vec{1} = 1$. Here $\mu_p = w^T \mu$ and $\sigma_p [w^T \Sigma w]^{1/2}$. Instead, we maximize $f(w) = (\mu_p - R_f)^2/\sigma_p^2$ subject to $g(w) = w^T \vec{1} = 1$, where as usual $\vec{1}$ is a vector of ones. Since

$$\nabla f(w) = \frac{2(\mu_p - R_f)(w^T \Sigma w)\mu - 2(\mu_p - R_f)^2 \Sigma w}{(w^T \Sigma w)^2}$$

and $\nabla g(w) = \vec{1}$, it follows that at the local extremum we have

$$(w^T \Sigma w)\mu - (\mu_p - R_f)\Sigma w = c\vec{1}$$

for some scalar c. Using the side condition, we get

$$c = cw^T \vec{1}(w^T \Sigma w)w^T \mu - \mu_p w^T \Sigma w + R_f w^T \Sigma w = R_f w^T \Sigma w$$

(using $w^T \mu = \mu_p$). Thus, at the extremum, the equation

$$(w^T \Sigma w)\mu - \mu_p \Sigma w + R_f \Sigma w = R_f (w^T \Sigma w)\vec{1}$$

holds, and so

$$(w^T \Sigma w)(\mu - R_f \vec{1}) = (\mu_p - R_f)\Sigma W.$$

Since $\mu_p > R_f$, it follows that w is a multiple of $\Sigma^{-1}(\mu - R_f \vec{1})$. Again using the side condition, we get

$$w = \frac{\Sigma^{-1}(\mu - R_f \vec{1})}{(\mu - R_f \vec{1})^T \Sigma^{-1} \vec{1}},$$

as claimed in **Section 5.2**.

Title: The Lagniappe Fund: The Story of Diversification and Asset Allocation

Solutions to Requirements

Requirement 1.

1a. Monthly returns are -0.0018, 0.0000, 0.0252, 0.0429, 0.0064, 0.0251, -0.0532, 0.0005, 0.0215, 0.0646, 0.0738, 0.0384.

1b. Mean of monthly returns = 0.0195, standard deviation of monthly returns = 0.0345. The annualized mean is 0.2342 and the annualized standard deviation is 0.1194.

1c. The monthly returns of the portfolio are 0.0115, 0.0297, 0.0626, 0.0484, 0.0119, -0.0440, -0.2067, 0.0176, -0.0228, 0.0384, 0.0609, 0.0116. The mean and standard deviation are 0.0171 and 0.0336. The annualized mean is 0.2050 and the annualized standard deviation is 0.1164.

Requirement 2.

2a. For the monthly returns, the mean and standard deviation are 0.0159 and 0.0313. The annualized mean is 0.1913 and the annualized standard deviation is 0.1083.

2b. For S&P500, EXXON, and FORD, the covariance matrix is

$$\begin{bmatrix} 9.77E-4 & 6.16E-4 & 9.09E-5 \\ 6.16E-4 & 1.19E-3 & 1.10E-8 \\ 9.09E-5 & 1.10E-8 & 3.33E-3 \end{bmatrix}.$$

The correlation matrix is

$$\begin{bmatrix} 1 & .5715 & .0504 \\ .5715 & 1 & .0000 \\ .0504 & .0000 & 1 \end{bmatrix}$$

2c. For the monthly returns, the mean is 0.01757 and the standard deviation is 0.0310. The annualized mean is 0.2109 and the annualized standard deviation is 0.1074.

2d. For the monthly returns, the mean is 0.0154 and the standard deviation is 0.0305. The annualized mean is 0.1857 and the annualized standard deviation is 0.1055.

2e. For the monthly returns, the mean is 0.0167 and the standard deviation is 0.02770. The annualized mean is 0.2005 and the annualized standard deviation is 0.0960.

Requirement 3. Using daily data, we have the results below.

	S&P	EXXON	FORD
Mean	0.00075	0.00097	0.00073
Standard deviation	0.00742	0.01230	0.01510
Annualized mean	0.1908	0.2454	0.1853
Annualized standard deviation	0.1182	0.1960	0.2406

The correlation matrix is

$$\begin{bmatrix} 1.0000 & 0.5411 & 0.3998 \\ 0.5411 & 1.0000 & 0.1480 \\ 0.3998 & 0.1480 & 1.0000 \end{bmatrix}.$$

For an equally weighted portfolio (one-third each) of the S&P 500, FORD, and EXXON: For the daily returns, the mean is 0.00082 and the standard deviation is 0.00868. The annualized mean is 0.2071 and the annualized standard deviation is 0.1383.

Requirement 4. The weights are .263 and .737 for FORD and EXXON respectively.

Requirement 5. The plot will be similar to **Figure 2**.

Requirement 6. To attain 2.5% expected monthly return, our weights should be 0.02/0.0145 = 1.379 for EXXON and −0.379 for the risk-free security. That is, we should take a short position in the security and put the cash received into EXXON. To attain a portfolio standard deviation of 0.025 for our monthly return, we should put weight 0.025/0.0344 = 0.725 into EXXON and weight 0.275 into the risk-free security. The plot will be similar to **Figure 4**.

Requirement 7. We can assume without loss of generality that $\sigma_1 > \sigma_2$. Letting $t = w_1$, the variance of the portfolio return is

$$v(t) = \sigma_1^2 t^2 + 2\rho_{12}\sigma_1\sigma_2 t(1-t) + \sigma_2^2(1-t)^2.$$

We calculate

$$v'(t) = 2[\sigma_1^2 - 2\rho_{12} + \sigma_2^2]t + \rho_{12}\sigma_1\sigma_2 - \sigma_2^2.$$

Setting $v'(t) = 0$ and solving for $t^{\min}$, we get

$$t^{\min} = w_1^{\min} = \frac{\sigma_2^2 - \rho_{12}\sigma_1\sigma_2}{\sigma_1^2 - 2\rho_{12}\sigma_1\sigma_2 + \sigma_2^2},$$

as required. The numerator and denominator of this expression are guaranteed to be positive as long as $\rho_{12} < \sigma_2/\sigma_1$. By a similar calculation,

$$w_2^{\min} = \frac{\sigma_1^2 - \rho_{12}\sigma_1\sigma_2}{\sigma_1^2 - 2\rho_{12}\sigma_1\sigma_2 + \sigma_2^2},$$

and $w_2^{\min} > 0$ if $\rho_{12} < \sigma_1/\sigma_2$. Since $\rho_{12} \le 1$ and $\sigma_1 > \sigma_2$, this condition is satisfied. Therefore, the weights on the minimum variance portfolio are positive when $\rho_{12} < \sigma_2/\sigma_1$.

Requirement 8. **Table S1** shows the weights in the S&P 500 Index, FORD, and EXXON, for a portfolio that has the lowest standard deviation for desired mean returns ranging from 0.015 to 0.020. The table also shows the standard deviation of the portfolio. The standard deviation-mean plot will be similar to **Figure 3**.

Table S1.
Answers to **Requirement 8**.

Mean	S&P 500	FORD	EXXON	Standard Deviation
0.015	0.9030	0.2648	−0.1679	0.0303
0.016	0.6755	0.2266	0.0979	0.0272
0.017	0.4479	0.1885	0.3637	0.0262
0.018	0.2203	0.1503	0.6294	0.0278
0.019	−0.0073	0.1121	0.8952	0.0314
0.020	−0.2349	0.0738	1.1609	0.0365

Requirement 9. For the monthly returns, the mean is 0.01637 and the standard deviation is 0.02771. The annualized expected return would be 0.1964 and the annualized standard deviation would be 0.09599.

Requirement 10. The plot will be similar to **Figure 6** with numerical values for the level of R_f and the tangency portfolio of FORD and EXXON stocks.

Requirement 11. The weights are 0.1919 for FORD, 0.8081 for EXXON. These result in monthly $\mu_T = 0.0186$ and $\sigma_T = 0.0299$. The annualized mean is 0.2232 and the annualized standard deviation is 0.1036.

Requirement 12. The weights are 0.1600 for FORD, 0.5616 for EXXON, and 0.2784 for the S&P500 Index fund. These result in monthly $\mu_T = 0.0177$ and $\sigma_T = 0.0271$. The annualized mean is 0.2124 and the annualized standard deviation is 0.0939. Adding a tangent from $(0, 0.005)$ to the plot in **Requirement 8** will result in a figure similar to **Figure 6**.

Requirement 13. People close to retirement should not be investing in portfolios that have a high variance, that is, those having high σ_p. This is because they are not able to wait for the market to recover and are less able to tolerate the risk. They should therefore be invested in the Bond Portfolio and have no assets in the Stock Portfolio. The letter to the potential retiree should indicate their current holdings in the Stock Portfolio and Bond Portfolio and recommend that the client reduce the amount invested in the Stock Portfolio over time. Of course, it is still the client's decision how to act on the recommendation.

Requirement 14. People who have a long career ahead of them and are far from retirement are better positioned to ride out the ups and downs of the market and are also better able to bear the risk of a loss. To maximize the potential return, they should be invested more in stocks. The letter should indicate the pros and cons of stock investment using some of the language in this ILAP. It should explain what is meant by standard deviation and how to maximize the expected return for each level of standard deviation (the Capital Market Line). It should also note that since the professor has many more years to retirement, the professor can tolerate more risk and should therefore consider investing a larger amount in the Stock Portfolio.

Requirement 15. Many of the Web pages of retirement funds carry material with much of the same ideas as the solutions to **Requirements 13–14**. Frequently, they also have a "quiz" that a potential investor can take to determine their risk tolerance—the level of standard deviation that they can tolerate. Based on the answers to the quiz, recommendations are made for the optimal mix of stocks and bonds.

Title: The Lagniappe Fund: The Story of Diversification and Asset Allocation

Notes for the Instructor

For this ILAP, students are assumed to have had

- calculus;

- an introduction to probability covering the basic properties of mean, variance, and standard deviation; and

- sufficient background in vectors and matrices to use matrix multiplication.

The more advanced aspects of linear algebra, such as eigenvalues and eigenvectors, are not needed.

This ILAP will probably work best in connection with a spreadsheet or other matrix-oriented computer package such as MATLAB or Mathematica.

The ILAP begins with the interpretation of mean and standard deviation as the level and risk of return of a portfolio. The students are required to calculate (estimate) these parameters for some simple portfolios. Next, the possibilities and benefits of diversification are explored, and the question of optimizing expected return subject to an acceptable level of risk is answered. Again, the students are expected to carry out some simple calculations reinforcing these basic financial concepts and the connection between elementary probability theory and theory of portfolios. The optimization of a portfolio in the presence of risk-free borrowing and lending is developed, including the theory of mean–variance efficiency and the efficient frontier; the students reinforce these developments with plots and calculations for simple portfolios. The role of the now famous β as a risk measure for financial assets is discussed (without calculations). The ILAP concludes with some practical suggestions for financial managers, based on the results in the previous sections of the ILAP.

This is an open-ended ILAP intended to emphasize the practical nature of the concepts discussed. If the students get really involved in this project, they may end up having more questions than answers! The instructor should indicate some of the references (at the end of the ILAP) for follow-up study.

Acknowledgments

We would like to thank COL David Arney (the editor), USMA, and LTC Michael Meese (the reviewer), USMA, for their feedback and valuable suggestions.

About the Authors

N.K. Chidambaran received his B.Tech. in chemical engineering from the Indian Institute of Technology, Bombay, and his Ph.D. in finance from New York University. He is Assistant Professor of Finance at the A.B. Freeman School of Business, Tulane University. His research interests are in the fields of corporate finance and financial derivatives. He has published his research in leading academic journals such as the *Journal of Financial Economics, Journal of Risk and Insurance, Journal of Derivatives*, and the *Financial Analysts Journal*. He has taught undergraduate and graduate courses in Financial Management, Financial Markets, Financial Derivatives, and Risk Management.

John Liukkonen received his B.A. in mathematics from Harvard University and Ph.D. in mathematics from Columbia University. Since 1970, he has been in the Mathematics Department at Tulane University, where he worked first in harmonic analysis and now works in statistics and statistics education.

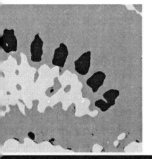

INTERDISCIPLINARY LIVELY APPLICATIONS PROJECT

UTHORS:
i Yu
ansportation Studies)
_lx@tsu.edu

ella D. Bell (Mathematics)
smddbell@tsu.edu

xas Southern University
ouston, TX

DITOR:
vid C. Arney

ONTENTS:

Travel Demand Forecasting and Analysis for South Texas

MATHEMATICS CLASSIFICATIONS:
Algebra, Calculus, and Statistics

DISCIPLINARY CLASSIFICATIONS:
Transportation Planning

PREREQUISITE SKILLS:
Solving linear equations
Solving quadratic equations
Regression analysis concept
Elementary optimization

PHYSICAL CONCEPTS EXAMINED:
Trip attractions, Trip distributions
Trip modal choice, Trip route choice

COMPUTER REQUIREMENTS:
Microsoft Excel or similar tool
Ability to use calculator

Contents

1.   Setting the Scene

Travel demand forecasting is the most important phase in the urban transportation planning process. The purpose of travel demand forecasting is to predict future travel demands on the roads in order to estimate the likely transportation consequences of transportation alternatives (including a do-nothing alternative) that are being considered for implementation. Usually, travel demand forecasting is performed using a four-step sequential model described as follows:

Step 1: Trip Generation: Should I make a trip?

Step 2: Trip Distribution: Where should I go?

Step 3: Modal Choice: What mode of transportation should I use?

Step 4: Trip Assignment: Which route in the network should I take?

Step 1 determines how many trips will be produced by each residential area and how many trips will be attracted by each commercial site. Step 2 determines where each trip generated in Step 1 will go. For example, for each residential area, there may exist multiple choices for the shopping purpose. The percentage of the total trips that will be attracted by each shopping site must be determined. Step 3 determines what mode of transportation each trip will use. The choices of transportation may include bus, light-rail (used only in the city), subway, private automobile, taxi, bicycle, walk, and so on. Finally, Step 4 determines which route each trip will use.

By the end of the travel-demand forecasting process, the traffic volumes (the number of vehicles per unit time) on the roads will be produced, which provide useful information about the congestion on the streets. Thereafter, transportation planners can select the best transportation projects by reviewing the resulting levels of congestion from a series of transportation alternatives. The four-step travel demand forecasting process is usually carried out using mathematical models.

2. Step 1: Trip Generation

General Information

The objective of a trip-generation model is to forecast the number of trips that will begin from or end in each travel zone within the region for a typical day of the target year. The most widely used trip generation models are regression models that are expressed by the following two linear equations:

$$P_i = p_0 + p_1 X_1 + p_2 X_2 + \cdots + p_r X_r,$$
$$A_i = a_0 + a_1 X_1 + a_2 X_2 + \cdots + a_r X_r,$$

where

P_i represents the total number of trips produced by a residential zone i;

A_i represents the total number of trips attracted by commercial activities in zone i;

$X_1, \ldots, X_r$ are independent variables, which are usually derived from the urban travel surveys. For example, X_1 may represent the total zonal population, X_2 may represent the average household income, and X_3 may represent the average auto ownership of each zone.

$p_0, \ldots, p_r, a_0 \ldots a_r$ are constant values/coefficients of the independent variables, which are usually derived through the regression analysis.

In many cases, the regression equations for trip productions and attractions can be nonlinear. However, since the regression analysis for nonlinear equations is often performed by converting the nonlinear form to a linear form first, understanding how to solve the linear equations is essential.

Example: Trip Attractions in South Texas

A survey of the commercial activities was conducted for five zones in South Texas. Each zone represents a portion of the land that is used as the basic unit in an urban planning process. The data were collected based on three types of employment: manufacturing, retail and services, and others. The resulted zonal employment of three different commercial types and their respective trip attractions are listed in **Table 1**.

- The first column is the zone number.

- The second column is the total number of people employed in manufacturing related companies for each zone.

- The third column is the total number of people employed in the retail and services sector.

- The fourth column is the total number of people employed in all companies other than manufacturing and retail and services.

- The fifth column is the total number of people employed for all companies.

- The last column lists the surveyed number of trips that are attracted by each zone.

Variables X_1, X_2, and X_3, totalling to X, are employment types, while the variable Y is trip attraction.

Table 1.

The number of people employed in each sector and the total trip attractions .

Zone	Manuf. X_1	Zonal Employment Ret & Ser X_2	Others X_3	Total X	Trip Attraction Y
1	6820	2547	115	9482	9428
2	111	1899	0	2010	2192
3	228	87	259	574	330
4	0	127	0	127	153
5	2729	813	294	3836	3948

This example helps exercise the concept of developing trip-attraction equations through the regression analysis technique. In practice, the attraction equations are used to predict the actual trip attractions given projected employment for a future year. Trip productions are predicted in the same way. To simplify the problem, this example is designed to practice just the regression analysis process, not to predict the future trip attractions. In Step 2 in the next section, trip productions and attractions are assumed to have been predicted beforehand.

Requirement 1: Determine a single linear regression equation between dependent variable Y and each of independent variables X, X_1, X_2, and X_3 individually.

Requirement 2 (Optional): Determine a multiple linear regression equation between the dependent variable Y and the independent variables X_1, X_2, and X_3. (Use of Microsoft Excel Data Analysis Tool is suggested.)

Requirement 3: Select the equations from **Requirements 1** and **2** that are acceptable for use in trip generation analysis. (A minimum value of R-squared, such as 0.65, should be set as the criterion and compared with the value of R-squared for each equation. Optimally, an F-test and/or t-test should be conducted.)

3. Step 2: Trip Distribution

The Gravity Model

The objective of a trip distribution model is to determine the total number of trips between all pairs of zones i and j, where i is the trip-producing zone and j is the trip-attracting zone of the pair. The rationale of trip distribution is as follows: All trip-attracting zones j in the region are in competition with each other to attract trips produced by each zone i. Everything else being equal, more trips will be attracted by zones that have higher levels of "attractiveness."

A widely used trip distribution model is the *gravity model*, which gets its name from being conceptually based on Newton's law of gravitation. That law states that the force of attraction between two bodies is directly proportional to the product of the masses of the two bodies and inversely proportional to the square of the distance between them:

$$F = k \frac{M_1 M_2}{r^2}.$$

The variation for trip distribution takes the following form:

$$T_{ij} = k \frac{P_i A_j}{W_{ij}^c}, \tag{1}$$

where

T_{ij} is the total number of trips between zones i and j;

P_i, A_j are the total trip productions for zone i and attractions for zone j, which can be derived from the trip generation step; A_j may also be the relative attractiveness of zone j;

k is a constant value for a given starting zone; and

W_{ij}^c is the travel impedance, which can be defined as either the travel time or the travel distance.

Rewriting **(1)** in a form that is expressed by the trip productions and a probability factor results in

$$T_{ij} = P_i \left(\frac{A_j F_{ij}}{\sum_x A_x F_{ix}} \right), \qquad \text{where} \quad F_{ij} = \frac{1}{W_{ij}^c}. \tag{2}$$

The quantity F_{ij} is called the *travel-time factor* or *friction factor*. The term contained by the parentheses is a probability factor, which represents the proportion that trips produced by zone i are attracted by zone j; this term is usually expressed by p_{ij}. Usually, the value of exponent c in **(2)** must be determined by calibration before the gravity model is applied.

Example: Trip Distribution in South Texas

In a small town in South Texas, the land uses are divided into four zones as shown in **Figure 1**.

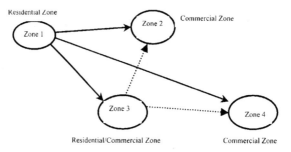

Figure 1. A four-zone network.

Zone 1 is residential, with no commercial sites. Zone 2 and Zone 4 are commercial, with no residential land uses. Zone 3 is a combined residential and commercial zone. Zones 2, 3, and 4 compete in attracting trips produced by the households in Zones 1 and 3.

Based on a trip-generation analysis, the total trip productions and the relative zonal attractiveness for the target-year 2005 are generated (**Table 2**); the inter-zonal and intra-zonal travel times are measured (**Table 3**). A calibration of the gravity model finds that $c = 2.0$ in **(2)**.

Table 2.

Trip productions P_i and attractiveness A_j for the four-zone network.

Zone	Productions	Attractiveness
1	20,000	0
2	0	3
3	15,000	2
4	0	6

Table 3.

Travel time W_{ij} between Zone i and Zone j.

$i \setminus j$	1	2	3	4
1	5	10	10	20
2	10	5	10	10
3	10	10	5	10
4	20	10	10	5

Requirement 4: Use the gravity model **(2)** to estimate the interchange trips between zones.

Requirement 5: Estimate the total trip attractions for Zones 2, 3, and 4.

4. Step 3: Modal Choice

General Information

In a typical travel situation, trip-makers can select from among several travel modes: driving, riding with someone else, taking the bus, walking, riding a motorcycle, and so forth. A *modal choice* (sometimes called *modal split*) is concerned with the trip-maker's behavior regarding the selection of travel mode.

The most widely used modal choice model is the *multinomial logit model*. The multinomial logit model calculates the proportion of travelers that will select a specific mode K according to the following relationship:

$$p(K) = \frac{e^{U_K}}{\sum_x e^{U_K}}, \tag{3}$$

where U_K is a utility function that measures the degree of satisfaction that people derive from their choices. The magnitude of U_K depends on the characteristics of each choice, which are called *attributes*, and on the characteristics of the individual making the choice, which are called *socioeconomic status*. The utility function is typically expressed as the linear weighted sum of the independent variables (or a transformation of them):

$$U_K = a_K + a_1 X_1 + a_2 X_2 + \cdots + a_r X_r,$$

where a_K is the calibrated mode-specific constant for the mode K and X_i is the ith attribute weighted by the model parameter a_i. The attributes may include such variables as the access and egress time, waiting time, line-haul time (actual time spent in the vehicle), out-of-pocket cost, level of service, and convenience.

Example: Modal Choice in South Texas

In the same town as used earlier, people who go from Zone 1 to Zone 4 have two choices of transportation: private automobile and local bus (which is operated by a local public transportation authority). A calibration process has resulted in the following utility function for the auto and bus:

$$U_K = a_K - 0.05X_1 - 0.02X_2 - 0.025X_3 - 0.001X_4,$$

where

X_1 = waiting time, in minutes;

X_2 = line-haul time, in minutes;

X_3 = access time, in minutes;

X_4 = out-of-pocket cost, in cents.

The interchange trips between Zone 1 and Zone 4 for the target-year 2005 has been forecasted as described earlier. The target-year service attributes of the two competing modes have been estimated as in **Table 4**.

Table 4.
Target-year service attributes for automobile and local bus.

Attribute	X_1	X_2	X_3	X_4
Automobile	0	15	5	80
Local Bus	15	35	10	50

Requirement 6: Assuming that the calibrated mode-specific constants a_K are -0.14 for the automobile and -0.60 for the bus, use the logit model equation (3) to estimate the target-year market shares of the automobile and bus and the resulting fare revenue of the bus system.

Requirement 7: A proposal is submitted to build a subway transit system (ST) between Zone 1 and Zone 4 for this town. A study has projected that the service attributes of the proposed subway system will be:

$$X_1 = 5, \qquad X_2 = 30, \qquad X_3 = 10, \qquad X_4 = 100.$$

Based on professional experience, the mode-specific constant for the subway system is -0.45. Estimate the new market shares of three modes that will result from implementing the subway system proposal and the effect on the revenues of the public transportation authority, which operates both the local bus and the subway system.

Requirement 8: What other factors need to be considered to decide if a subway should be built? Explain.

5.   Step 4: Trip Assignment

General Information

The last step in the sequential travel demand forecasting process is concerned with the trip-makers' choice of path between pairs of zones by travel mode and with the resulting vehicular flows on the multimodal transportation network. The trip assignment problem can be described as follows: given traffic-demand departure rates for each Origin Destination (OD) pair in the network, which are usually derived from the Steps 1–3 of the travel demand forecasting process, determine the likely path choices and path-use probabilities for each OD pair in the network, and predict the resulting traffic flows and link travel times on each of the individual links that make up the network.

Most of the methods for solving the trip assignment problem are based on two trip assignment principles, which were first presented by Wardrop [1952]. These principles can be stated as follows:

- **Wardrop's 1st Assignment Principle:** Each driver chooses a path that minimizes his own travel time through the network.

- **Wardrop's 2nd Assignment Principle:** Each driver chooses a path that minimizes the total network travel time of all drivers.

When the network trips are assigned so that no driver can reduce his own travel time by switching from his current route to an alternative, the traffic flows are said to be in a *user-equilibrium state*. This state coincides with Wardrop's 1st principle. Similarly, when the network trips are assigned so that no driver can reduce the total network travel time by switching from his current route to an alternative, the traffic flows are said to be in a *system equilibrium state*. This state coincides with Wardrop's 2nd principle.

The trip assignment problems can be formulated as nonlinear mathematical programs based on Wardrop's principles. We will use a simple two-node and two-link network as an example. We assume a network with only two nodes and two links, as shown in **Figure 2**.

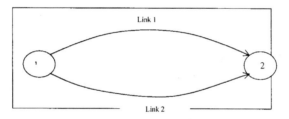

Figure 2. Illustration of a simple 2-node 2-link network.

Node 1 is an origin, where the trips generate, and Node 2 is a destination, where the trips sink. There are two routes available for driving from Node 1 to Node 2, through either Link 1 or Link 2. The link travel time on either Link 1 or Link 2 is a function of the traffic flow on the link. In other words, link travel time will vary with the change of the number of vehicles on the link. If it is assumed that r_{12} vehicles per hour go from Node 1 to Node 2, the objective of the trip assignment process is to determine how much traffic of the r_{12} will use Link 1 and how much will use Link 2.

The link travel time functions are expressed as

$$tt_1 = f(X_1), \tag{4}$$

$$tt_2 = f(X_2), \tag{5}$$

where X_1 and X_2 are the total number of vehicles per hour on Link 1 and Link 2, and tt_1 and tt_2 are the travel times on Link 1 and Link 2. To perform the trip assignment for this simple network, two formulations can be made, each based on one of Wardrop's principles.

Based on Wardrop's 1st Assignment Principle

The travel times on Link 1 and Link 2 must be equal after all trips are assigned. Otherwise, some vehicles on the higher travel time link will automatically switch to the lower travel time link, as each individual driver seeks to minimize his/her own travel times. Based on this logic and the assumption that r_{12} vehicles per hour go from Node 1 to Node 2, the following equations hold:

$$f(X_1) = f(X_2), \tag{6}$$
$$X_1 + X_2 = r_{12}. \tag{7}$$

These equations contain only two variables. Therefore, X_1 and X_2 can often be solved for using these two simultaneous equations if $f(X_1)$ and $f(X_2)$ are linear equations. Accordingly, the link travel times can also be calculated using (4) or (5).

Based on Wardrop's 2nd Assignment Principle

The total travel times of Link 1 and Link 2 should be minimum. Otherwise, some vehicles will switch the link in order to reduce the total travel times. The total travel times for Link 1 and Link 2 are expressed as follows:

$$TT = X_1 f(X_1) + X_2 f(X_2). \tag{8}$$

Therefore, the following optimization formulation can be derived:

$$\begin{aligned} \text{minimize} \quad & X_1 f(X_1) + X_2 f(X_2) \\ \text{subject to} \quad & X_1 + X_2 = r_{12}. \end{aligned} \tag{9}$$

The solution of (9) will result in the values of X_1 and X_2. Thereafter, the individual link travel times as well as the total link travel times can be calculated based on (4), (5), and (8).

Example: Route Choice in South Texas

In the same example network as earlier, it is assumed that the automobile users driving from Zone 1 to Zone 4 have two available routes to use, similar to what has been illustrated in **Figure 2**. During the morning peak hours, the travel times using the two routes are functions of the vehicles using each route, as indicated by (4) and (5). We assume that travel time functions for Route 1 and Route 2 are expressed by the following equations:

$$tt_1 = 10 + 0.02 X_1,$$
$$tt_2 = 15 + 0.01 X_2,$$

where tt_1 and tt_2 are the actual travel times (in minutes) on Route 1 and Route 2 and X_1 and X_2 are the number of vehicles using Route 1 and Route 2 per hour. It is estimated from the trip distribution and modal split process that during the morning peak hours, about 2000 vehicles per hour will leave Zone 1 heading for Zone 4.

Requirement 9: Based on Wardrop's 1st (user equilibrium) principle and **(6)** and **(7)**, formulate the trip assignment problem. Solve the formulation and estimate the number of vehicles that will use Route 1 and the number of vehicles that will use Route 2. Calculate the resulting travel times on each route and the total vehicle travel times between Zone 1 and Zone 4.

Requirement 10: Based on Wardrop's 2nd (system optimal) principle and **(8)**, formulate the trip assignment problem. Solve the formulation and estimate the number of vehicles that will use Route 1 and the number of vehicles that will use Route 2. Calculate the resulting travel times on each route and the total vehicle travel times between Zone 1 and Zone 4.

Requirement 11: Compare the results from **Requirement 9** and **Requirement 10**. Check if these results are consistent with the Wardrop's trip assignment principles. Comment on your findings.

References

Wardrop, J.G. 1952. Some theoretical aspects of road traffic research. *Proceedings of the Institute of Civil Engineering* Part II, 1: 325–378.

Title: **Travel Demand Forecasting and Analysis for South Texas**

Sample Solution

Step 1: Trip Generation

Requirement 1: Using the Excel Tool⟶Data Analysis⟶Regression, the linear regressions for Y vs. X, Y vs. X_1, Y vs. X_2, and Y vs. X_3 result in the output in **Tables 5–8**.

Table 5. Simple linear regression result for trip attraction and total employment.

SUMMARY OUTPUT

Regression Statistics	
Multiple R	0.999061
R Square	0.998124
Adjusted R Square	0.997498
Standard Error	190.2617
Observations	5

ANOVA

	df	SS	MS	F	Significance F
Regression	1	57775542	57775542	1596.03	3.45E-05
Residual	3	108598.6	36199.53		
Total	4	57884141			

	Coefficients	Standard Error	t Stat	P-value	Lower 95%	Upper 95%	Lower 95.0%	Upper 95.0%
Intercept	0.163903	117.0305	0.001401	0.99897	-372.28	372.6074	-372.28	372.60
Total	1.001321	0.025064	39.95035	3.45E-05	0.921556	1.081087	0.921556	1.0810

Table 6. Simple linear regression result for trip attraction and manufacturing employment.

SUMMARY OUTPUT

Regression Statistics	
Multiple R	0.977187
R Square	0.954894
Adjusted R Square	0.939859
Standard Error	932.9017
Observations	5

ANOVA

	df	SS	MS	F	Significance F
Regression	1	55273224	55273224	63.51013	0.004122
Residual	3	2610917	870305.6		
Total	4	57884141			

	Coefficients	Standard Error	t Stat	P-value	Lower 95%	Upper 95%	Lower 95.0%	Upper 95.0%
Intercept	705.9511	522.3078	1.3516	0.269385	-956.267	2368.169	-956.267	2368.1
Manuf.	1.266307	0.158898	7.969325	0.004122	0.760623	1.771991	0.760623	1.7719

Table 7. Simple linear regression result for trip attraction and retail and service employment.

SUMMARY OUTPUT

Regression Statistics	
Multiple R	0.832851
R Square	0.69364
Adjusted R Square	0.59152
Standard Error	2431.28
Observations	5

ANOVA

	df	SS	MS	F	Significance F
Regression	1	40150774	40150774	6.792412	0.079945
Residual	3	17733367	5911122		
Total	4	57884141			

	Coefficients	Standard Error	t Stat	P-value	Lower 95%	Upper 95%	Lower 95.0%	Upper 95.0%
Intercept	40.22623	1631.45	0.024657	0.981877	-5151.78	5232.234	-5151.78	5232.2
Ret&Ser	2.896011	1.11119	2.606226	0.079945	-0.64029	6.432316	-0.64029	6.4323

Table 8. Simple linear regression result for trip attraction and other employment.

SUMMARY OUTPUT

Regression Statistics	
Multiple R	0.087814
R Square	0.007711
Adjusted R Square	-0.32305
Standard Error	4375.606
Observations	5

ANOVA

	df	SS	MS	F	Significance F
Regression	1	446360.4	446360.4	0.023314	0.888336
Residual	3	57437780	19145927		
Total	4	57884141			

	Coefficients	Standard Error	t Stat	P-value	Lower 95%	Upper 95%	Lower 95.0%	Upper 95.0%
Intercept	2889.568	2870.336	1.006701	0.388241	-6245.13	12024.27	-6245.13	12024
Others	2.399937	15.71792	0.152688	0.888336	-47.6216	52.42143	-47.6216	52.421

Requirement 2: Using the Excel Tool⟶Data Analysis⟶Regression, the linear regression for Y vs. X_1, X_2, and X_3 results in the output in **Tables 9–10.**

Requirement 3: The following equations have $R^2 \geq .65$:

$$Y = 0.163903 + 1.001321X, \qquad\qquad R^2 = .998;$$

$$Y = 705.9511 + 1.266307X_1, \qquad\qquad R^2 = .955;$$

$$Y = 40.22623 \qquad\qquad +2.896011X_2, \quad R^2 = .694;$$

$$Y = 119.7625 \qquad + 1.001821X_1 \quad +1.013371X_2, \quad R^2 = .998.$$

If F-tests and t-tests are conducted to further examine the validity of each equation, all but the third equation are found to be statistically acceptable.

Table 9. Simple linear regression result for trip attraction and retail and service employment.

SUMMARY OUTPUT

Regression Statistics	
Multiple R	0.99938
R Square	0.998761
Adjusted R Square	0.995043
Standard Error	267.8365
Observations	5

ANOVA

	df	SS	MS	F	Significance F
Regression	3	57812404	19270801	268.6336	0.044814
Residual	1	71736.39	71736.39		
Total	4	57884141			

	Coefficients	Standard Error	t Stat	P-value	Lower 95%	Upper 95%	Lower 95.0%	Upper 95.0%
Intercept	-32.1931	284.1409	-0.1133	0.928177	-3642.53	3578.144	-3642.53	3578.1
Manuf.	0.969715	0.079029	12.27035	0.051768	-0.03444	1.97387	-0.03444	1.973
Ret&Ser	1.105403	0.217178	5.089861	0.123503	-1.65409	3.864894	-1.65409	3.8648
Others	0.858613	1.243212	0.690641	0.615216	-14.9378	16.65505	-14.9378	16.655

Table 10. Simple linear regression result for trip attraction and other employment.

SUMMARY OUTPUT

Regression Statistics	
Multiple R	0.999084
R Square	0.99817
Adjusted R Square	0.996339
Standard Error	230.1669
Observations	5

ANOVA

	df	SS	MS	F	Significance F
Regression	2	57778187	28889094	545.3159	0.00183
Residual	2	105953.6	52976.8		
Total	4	57884141			

	Coefficients	Standard Error	t Stat	P-value	Lower 95%	Upper 95%	Lower 95.0%	Upper 95.0%
Intercept	119.7625	154.5094	0.775115	0.519369	-545.0381	784.563	-545.038	784.5
Manuf.	1.001821	0.054921	18.24112	0.002992	0.765515	1.238128	0.765515	1.238'
Ret&Ser	1.013371	0.14737	6.876347	0.020501	0.379286	1.647455	0.379286	1.647<

Step 2: Trip Distributions in South Texas

Requirement 4: For the origin $i = 1$, the trip production $P_1 = 20000$. Based on the **(2)** and **Table 2**, the friction factor F_{1j} can be calculated as follows:

$$F_{11} = \frac{1}{W_{11}^2} = \frac{1}{5^2} = 0.04, \qquad F_{12} = \frac{1}{W_{12}^2} = \frac{1}{10^2} = 0.01,$$

$$F_{13} = \frac{1}{W_{13}^2} = \frac{1}{10^2} = 0.01, \qquad F_{14} = \frac{1}{W_{14}^2} = \frac{1}{20^2} = 0.0025.$$

The interchange trips between Zone 1 and other zones can then be calculated based on **(2)**. To simplify the calculation process, we use **Table S1** to perform the calculation.

Table S1.

j	A_j	F_{1j}	$A_j F_{1j}$	P_{1j}	T_{1j}
1	0	0.04	0	0	0
2	3	0.01	0.03	0.462	9240
3	2	0.01	0.02	0.308	6160
4	6	0.0025	0.015	0.230	4600
		SUM =	0.065	1.000	

For the origin $i = 3$, the trip production $P_3 = 15000$. A similar calculation in **Table S2** can be used to calculate the interchange trips between Zone 3 and other zones.

Table S2.

j	A_j	F_{3j}	$A_j F_{3j}$	P_{3j}	T_{3j}
1	0	0.01	0	0	0
2	3	0.01	0.03	0.176	2640
3	2	0.04	0.08	0.471	7065
4	6	0.01	0.06	0.353	5295
		SUM =	0.17	1.000	

For origins 2 and 4, since there are no trip productions, the interchange trips between these two zones and other zones become zero. Therefore, the final interchange trips can be summarized into the **Table S3**.

Requirement 5: If a row in **Table S3** is summed, the sum should equal the total trip productions for that zone. On the other hand, if a column in the table is summed, the total trip attractions for that can be derived. Therefore, the total trip attractions for Zones 2, 3, and 4 are calculated as follows:

Table S3.

$i \setminus j$	1	2	3	4
1	0	9240	6160	4600
2	0	0	0	0
3	0	2640	7065	5295
4	0	0	0	0

$$A_2^* = 9240 + 2640 = 11880,$$
$$A_3^* = 6160 + 7065 = 13225,$$
$$A_4^* = 4600 + 5295 = 9895.$$

Step 3: Modal Choice

Requirement 6: Based on the trip distribution analysis in Step 2, the interchange trips between Zone 1 and Zone 4 are 4600. Given the utility function and the attribute values by Table 3, the utilities for automobile and bus can be calculated as

$$U(\text{auto}) = -0.14 - 0.05(0) \ -0.02(15) - 0.025(5) - 0.001(80) = -1.095,$$
$$U(\text{bus}) = -0.60 - 0.05(15) - 0.02(35) - 0.25(10) - 0.001(50) = -2.35.$$

Therefore, the proportions of trips that will select the automobile and bus can be estimated as follows using **(3)**:

$$p(\text{auto}) = \frac{e^{-1.095}}{e^{-1.095} + e^{-2.35}} = 78\%,$$
$$p(\text{bus}) = \frac{e^{-2.35}}{e^{-1.095} + e^{-2.35}} = 22\%.$$

The market shares of the automobile and bus are calculated as

$$T_{14}(\text{auto}) = (0.78)(4600) = 3500 \text{ trips/day},$$
$$T_{14}(\text{bus}) = (0.22)(4600) = 1012 \text{ trips/day}.$$

The fare revenue of the bus system is (1012 trips/day)($0.50/trip) = $506/day.

Requirement 7: By introducing the subway system, the utilities for automobile and bus will stay the same as those calculated in the above **Requirement 6** which are $U(\text{auto}) = -1.095$ and $U(\text{bus}) = -2.35$. The utility of the proposed subway system is calculated as follows:

$$U(\text{subway}) = -0.45 - 0.05(5) - 0.02(30) - 0.025(10) - 0.001(100) = -1.65.$$

Therefore, the proportions of trips that will use each mode are estimated as:

$$p(\text{auto}) = \frac{e^{-1.095}}{e^{-1.095} + e^{-2.35} + e^{-1.65}} = 54\%,$$

$$p(\text{bus}) = \frac{e^{-2.35}}{e^{-1.095} + e^{-2.35} + e^{-1.65}} = 15\%,$$

$$p(\text{subway}) = \frac{+e^{-1.65}}{e^{-1.095} + e^{-2.35} + e^{-1.65}} = 31\%.$$

The new market shares after the introduction of the subway system are calculated as follows:

$$T_{14}(\text{auto}) = (0.54)(4600) = 2484 \text{ trips/day},$$
$$T_{14}(\text{bus}) = (0.15)(4600) = 690 \text{ trips/day},$$
$$T_{14}(\text{subway}) = (0.31)(4600) = 1426 \text{ trips/day}.$$

The new fare revenue of both bus and subway systems is

(690 bus trips/day)($0.50/bus trip)
    + (1426 subway trips/day)($1.00/subway trip) = $1,771/day.

The revenue increase of the public transportation authority (which operates both the local bus and subway systems) due to the introduction of the subway system is $1,265 (= $1,771 − $506). The increase in fare revenue alone does not justify the construction of the subway; a comprehensive cost-benefit analysis should be conducted before the decision is made.

Requirement 8: Free thoughts.

Step 4: Trip Assignment

Requirement 9: Based on Wardrop's 1st principle, **(6)** and **(7)**, and the travel time functions for Route 1 and Route 2, the following linear simultaneous equations can be established:

$$10 + 0.02X_1 = 15 + 0.01X_2,$$
$$X_1 + X_2 = 2000.$$

These equations can be solved for $X_1 = 833.33$ and $X_2 = 1166.67$, which means that 833.33 vehicles will actually use Route 1 and 1166.67 vehicles will use Route 2. The travel times are

Travel time on Route 1 = 10 + 0.02(833.33) = 26.67 min
Travel time on Route 2 = 15 + 0.01(1166.67) = 26.67 min
Total travel time = (833.33)(26.67) + (1166.67)(26.67) = 53,340 vehicle-min

Requirement 10: Based on Wardrop's 2nd principle, **(8)**, and the travel-time functions, the following optimization formulation can be established:

$$\text{minimize} \quad TT = X_1(10 + 0.02X_1) + X_2(15 + 0.01X_2)$$
$$\text{subject to} \quad X_1 + X_2 = 2000,$$

where TT is the total system travel time for all vehicles going from Zone 1 to Zone 4. The constraint can be rewritten as $X_2 = 2000 - X_1$. Substituting this into the objective function results in

$$TT = X_1(10 + 0.02X_1) + (2000 - X_1)[15 + 0.01(2000 - X_1)]$$
$$= 0.03X_1^2 - 45X_1 + 70000.$$

By setting the derivative of the above equation to zero, the optimal value for X_1 can be found:

$$\frac{dTT}{dX_1} = (0.03)(2X_1) - 45 = 0,$$
$$X_1 = 750,$$
$$X_2 = 2000 - 750 = 1250.$$

Therefore, based on Wardrop's second assignment principle, 750 vehicles will use Route 1 and 1250 vehicles will use Route 2. The travel times are calculated as follows:

Travel time on Route 1 = 10 + 0.02(750) = 25 min
Travel time on Route 2 = 15 + 0.01(1250) = 27.5 min
Total travel time = (750)(25) + (1250)(27.5) = 53,125 vehicle-min

Requirement 11: Based on Wardrop's 1st principle, all the used routes should have equal travel times. In **Requirement 9**, both Routes 1 and 2 have a travel time of 26.67 minutes, so the result conforms to Wardrop's 1st principle. In **Requirement 10**, the two routes have different travel times, so the result does not conform to the Wardrop's 1st assignment principle.

Based on Wardrop's 2nd principle, the total travel time should be a minimum. **Requirement 10** resulted in total travel time of 53,125 vehicle-minutes, which is less than the total travel time of 53,340 in **Requirement 9**. This means that although Wardrop's 2nd assignment principle does not result in equal travel times for all routes, it does result in minimum total travel times.

Title: Travel Demand Forecasting and Analysis for South Texas

Notes for the Instructor

This project is designed to exercise some basic concepts in regression analysis, algebra, calculus, and optimization theory, using a realistic example of transportation planning process. Understanding the travel demand forecasting process is critical in this project, while the mathematical calculation part for each assignment is not very difficult. The project is suitable for either individual or group work. With the understanding of the travel demand forecasting process concept by the students, the instructor can expand the assignment in each part to let students practice. The assignment in each part can also be used independently for exercise of mathematical topics. The requirements for Step 1 through Step 4 can also be used for a complete project for students. Some additional suggestions are as follows:

- In Step 1, similar problems can be also designed for trip productions in which the zonal population, household income, and the household automobile ownership become the independent variables while the trip production becomes the dependent variable. If a single independent variable is used, students can be asked to do the simple linear regression analysis manually. Nonlinear regression equations can be considered to let students practice how to transform a nonlinear equation to a linear equation.

- In Step 2, variations of the requirement can be designed:

    - More zones can be added to the analysis areas.

    - Residential zones and commercial zones can be redesigned.

    - The attractiveness can be assigned different values to test how the changes will affect the trip distribution results.

    - The travel time table can use different values for each cell.

    - The exponent c in the gravity equation can be assigned a different value other than 2, to examine how different values of c will change people's choices of commercial zones.

- In Step 3, students can be asked to change the attribute values for automobile and local bus—for example, the out-of-pocket cost or waiting time—to examine how the changes will affect the choice of mode. The coefficients of attributes in the utility function can also be redesigned so that a separate problem can be created. In addition to a subway system, more modes of travel can be considered, such as motorcycling, walking, ride-sharing, and so on, in which case values of various attributes for each added mode need to be assumed.

- In Step 4, three-route and four-route problems can be designed, which will make the mathematical equations more complicated. In designing additional route(s) to the network, new travel time function(s) should be assumed. Essentially, different forms of the travel time functions reflect different conditions of the roads. For example, some roads have one lane and some roads have two lanes in each direction. Some roads may have a lot of parked vehicles on the roadside, which significantly impede the moving of the vehicles, while some roads may have uphill and downhill slopes or short radius on curves. The travel time functions can also be made nonlinear, which will make the optimization more complicated.

About the Authors

Lei Yu is Associate Professor and Chair of the Transportation Studies Department at Texas Southern University. He is also a Changjiang Scholar of Northern Jiaotong University, China. He received a bachelor's degree in transportation engineering from Northern Jiaotong University, Beijing, China and a master's degree in production and systems engineering from Nagoya Institute of Technology, Nagoya, Japan. His Ph.D. degree in civil engineering was awarded by Queen's University in Kingston, Ontario, Canada.

As a professor at Texas Southern University, he has been teaching the courses in Highway Traffic Operations, Travel Demand Forecasting and Analysis, Transportation Design and Engineering, Computer Applications in Transportation, and Quantitative Analysis in Transportation. His research interests and expertise involve transportation network modeling, ITS related technologies and applications, dynamic traffic assignment and simulation, vehicle exhaust emission modeling, highway traffic control and operation strategies, travel demand forecasting models, and air quality issues in transportation.

In the past, Yu has served as the Principal Investigator (PI) of more than 20 research and consulting projects that were sponsored by various agencies. He has published numerous research papers in scientific journals and conference proceedings, plus project reports. Professionally, Dr. Yu is a licensed professional engineer (P.E.) in Texas, an active member of the Institute of Transportation Engineers (ITE), the American Society of Civil Engineers (ASCE), and the Transportation Research Board (TRB). He also is a member of numerous committees, councils, and task forces in regional, state, national and international organizations.

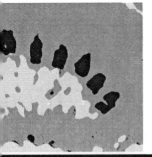

INTERDISCIPLINARY LIVELY APPLICATIONS PROJECT

THORS:
ye I. Selco (Chemistry)
ye_selco@redlands.edu

et L. Beery (Mathematics)
et_beery@redlands.edu

iversity of Redlands
dlands, CA

ITOR:
vid C. Arney

Pollution Police

MATHEMATICS CLASSIFICATIONS:
Linear Algebra, Abstract Algebra

DISCIPLINARY CLASSIFICATIONS:
Physical Chemistry (spectroscopy)

PREREQUISITE SKILLS:
Matrix representations of linear transformations
Group theory through classifications of groups
of small order (up to 12)

PHYSICAL CONCEPTS EXAMINED:
Molecular vibrations, symmetries, centers of mass,
and spectroscopy

COMPUTING REQUIREMENT:
Molecular modeling program

EQUIPMENT REQUIREMENT:
Ball-and-stick models of molecules
Infrared spectrometer

Contents

Setting the Scene

You have been hired by an environmental testing company to monitor air pollutants. Some of these chemicals are released into the atmosphere from factories, cars, or cattle; others evaporate from agricultural fields.

Since most of these chemicals absorb infrared light, we can detect them with an infrared spectrometer. To sample the air for pollutants, you focus the infrared light from a portable spectrometer onto a mirror (a few centimeters to 1 km away) and collect the light reflected back into the spectrometer. The spectrometer uses a prism to separate the infrared frequencies; the prism is turned slowly, so that at any given time the detector sees light only of a specific frequency (wavelength). The output from the spectrometer, an *infrared spectrum*, is a plot of the intensity of light reaching the detector divided by the initial intensity as a function of frequency, that is, $T = I/I_0$ vs. frequency in wavenumbers. The light reflected back to the spectrometer is light that was *not* absorbed by molecules in the atmosphere. Therefore, the resulting spectrum has downward going *peaks* when an absorption occurs at a given frequency (see **Figure 1**). The goal of this project is to identify atmospheric pollutants from their infrared spectra.

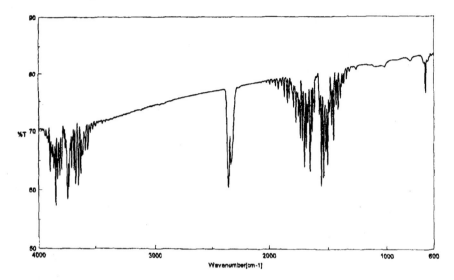

Figure 1. An infrared spectrum of air; this spectrum is due primarily to carbon dioxide and water.

Some of the molecules commonly found in the atmosphere are listed below. Water and small amounts of carbon dioxide, methane, and sulfur trioxide are found in "clean" air samples. Large amounts of any of the chemicals other than water usually indicate atmospheric pollution.

H_2O (water): Ubiquitous in the atmosphere, even on dry days.

CO_2 (carbon dioxide): Produced by the combustion of fuels (including the food we eat) and a major source of greenhouse warming in the atmosphere.

CH_3Br (methyl bromide): A pesticide that is sprayed on agricultural fields. Currently, it is under strict EPA controls; it is partially responsible for the ozone hole.

CH_4 (methane): Used primarily as natural gas; also produced by cows and rice fields.

CCl_2F_2 (dichlorodifluoromethane): A refrigerant that is partially responsible for the ozone hole.

C_2H_4 (ethylene): One of the components of natural gas; used to ripen bananas.

C_3H_4 (allene): Found in interstellar dust clouds; an explosive chemical waste.

SO_3 (sulfur trioxide): Results from the burning of sulfur containing fuels such as coal; reacts with water to form acid rain.

C_6H_6 **(benzene):** A component of gasoline; carcinogenic.

$C_6H_5CH_3$ **(toluene):** Also a component of gasoline; it is the standard compound used for measuring octane ratings with an octane rating of 100.

SF_6 **(sulfur hexafluoride):** Its presence is often an indication of nuclear weapons development, since it is a surrogate compound for uranium hexafluoride (UF_6); UF_6 is used to separate uranium isotopes to produce "enriched" uranium. Sulfur hexafluoride also is used as a gas insulator for high-voltage equipment.

Compare three-dimensional ball-and-stick models for these molecules with the two-dimensional representations in **Figure 2**, in which

- straight lines represent bonds that lie in the plane of the paper,

- two lines between a pair of atoms represent a double bond,

- a filled arrowhead indicates a single bond that is angled out of the plane of the paper toward you, and

- the dashed arrowhead indicates a bond angled into the plane of the paper away from you.

Carbon dioxide is an example of a *linear* molecule; water, ethylene, sulfur trioxide, and benzene are *planar* molecules.

In the case of toluene, the methyl group (CH_3) at the top spins freely and is called a "free rotor." This means that the three hydrogen atoms are not locked into place, as they have been drawn, but act as a freely spinning pinwheel. In fact, they can be modeled as a single atom having three times the mass of a hydrogen atom at the appropriate geometrical distance from the carbon; toluene then can be viewed as a planar molecule.

Requirement 1: Center of Mass and Coordinate Axes

To accomplish the goal of identifying a chemical from its infrared spectrum, we must know which infrared frequencies a molecule absorbs. To do this, we examine various properties of the molecules themselves, including their centers of mass, their Cartesian coordinate axes, their vibrations, and their symmetries. Specifically, we compare the actions of the symmetry operations of a molecule on its Cartesian coordinate axes with the actions of the symmetry operations on its vibrations.

When a molecule absorbs infrared radiation of a given frequency, this energy causes the molecule to vibrate in a specific way; the atoms bounce against each other much like balls connected by a spring. The vibrational motions of a

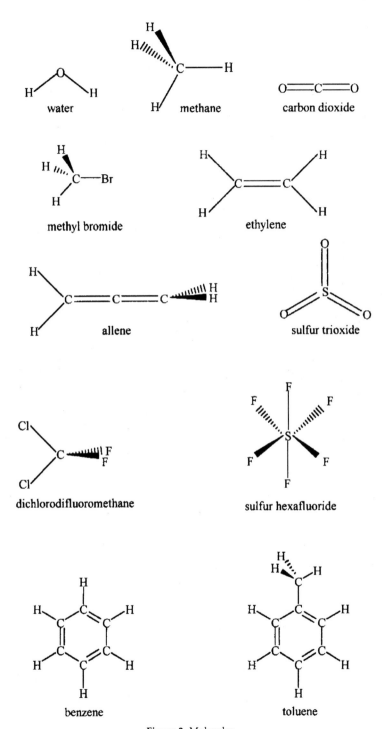

Figure 2. Molecules.

molecule that absorb infrared radiation exhibit the same behavior as do the Cartesian coordinate axes of the molecule when the atoms of the molecule are permuted in certain ways. This is a result of the orthogonal interaction between the electromagnetic field of the light and the electric field of the molecule itself.

Therefore, begin by sketching in a three-dimensional coordinate system for each of the molecules that your instructor assigns to you. The convention for molecules is that *the origin of the axis system is placed at the center of mass of the molecule*. (The center of mass is the weighted average of the masses of the atoms.) First, determine approximately where this should be. (You might want to re-examine the ball-and-stick models.) Remember, the masses of the different atoms are different. To find out how much each atom weighs, consult a periodic table of the elements. The mass number for each type of atom appears at the bottom of the square in which the atomic symbol appears.

By convention, the z-axis is the unique axis, if there is one. This axis is also called the *molecular axis*, or *axis of highest symmetry*. In a linear molecule, such as CO_2, it corresponds to the line formed by the molecule. For benzene, the z-axis is the out-of-plane axis, since that is the unique axis. If there doesn't seem to be a unique axis, then place the heaviest atoms in the molecule along the z-axis (often, there is more than one way to do this). *Once the z-axis is assigned, the in-plane axis usually is the y-axis and the out-of-plane axis is the x-axis.* In addition, axes should be placed along the molecular bonds whenever possible.

Example 1: The y- and z-axes for water and carbon dioxide are shown in **Figure 3**. The x-axis is out-of-plane from the origin (just below the O atom and pointing straight out at you for water, but centered in the C atom and pointing straight out at you for carbon dioxide).

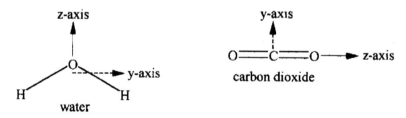

Figure 3. Cartesian coordinate axes for water and carbon dioxide.

Draw in the three Cartesian coordinate axes in a picture of each molecule as outlined in the steps below. (If you do not assign them correctly now, you will have a chance later to relabel them.)

- Estimate, by eye, the center of mass for each molecule, keeping in mind the atomic masses for each type of atom.

- Draw in the z-axis using the rules above.

- Draw in the two remaining axes using the rules given above.

Requirement 2: Molecular Vibrations

The types of molecular vibrations that a molecule has determine whether or not it absorbs infrared light. Hence, you need to determine the types of vibrations that your molecules make.

To ensure that all of the vibrations have been identified, we need to know how many are possible. Consider a molecule that is a collection of N atoms connected together in a specific way by chemical bonds. To describe the motions of the molecule, we need to consider the motions of each individual atom. This means that we need 3 degrees of freedom for every atom within the molecule, for a total of $3N$ degrees of freedom.

However, the atoms within the molecule have a specific geometric relationship to the other atoms in the molecule; this results in a reduction of the number of independent degrees of freedom. To describe the motion through space of the molecule, we must use three degrees of freedom, reducing the $3N$ degrees of freedom to $3N - 3$. Since the molecule also can rotate (like a spinning baton or Frisbee), we require 2 more degrees of freedom to describe the coordinates about which a linear molecule can spin and 3 more degrees of freedom for a nonlinear molecule. (For a linear molecule, there is no concerted rotation about the molecular axis (z-axis).) This leaves $3N - 5$ degrees of freedom for the linear molecule and $3N - 6$ for the nonlinear molecule still unaccounted for; each of the remaining degrees of freedom describes a distinct coordinated internal motion, or vibration, of the atoms within the molecule.

For example, when there are only two atoms in the molecule (e.g., oxygen (O_2), nitrogen (N_2), or carbon monoxide (CO), there is only one vibrational motion: $3(2) - 5 = 1$. In the case of benzene (C_6H_6), there are 12 atoms and $3(12) - 6 = 30$ vibrational motions possible! As it turns out, not all of these vibrations are capable of absorbing infrared radiation. For the simplest molecules, such as water, it is easy to draw pictures representing the vibrational motions.

Example 2: Water has $3(3) - 6 = 3$ vibrations and carbon dioxide has $3(3) - 5 = 4$ vibrations, as shown in **Figures 4** and **5**.

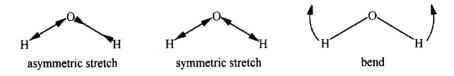

asymmetric stretch   symmetric stretch   bend

Figure 4. Vibrations of water.

Consider the water molecule as it undergoes the asymmetric stretch; as it reaches its most extreme position, it has one "arm" extended and the other "arm" contracted.

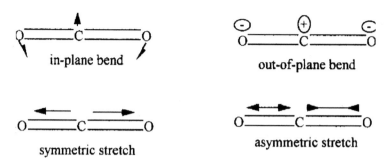

Figure 5. Vibrations of carbon dioxide.

For the out-of-plane bending motion of carbon dioxide, the + indicates that the motion of the atom is directly out of the plane of the paper towards you, while the − indicates that the motion of the atom is directly out of the plane of the paper away from you.

Record the motions and frequencies of the molecular vibrations as outlined in the steps below:

- For each molecule, determine the number of vibrational motions, $3N - 6$ or $3N - 5$, so you know how many vibrational motions to record.

- Use a computer program to view the vibrational motions for water, carbon dioxide, and the other molecules assigned to you by your instructor. Record the motions of the atoms in the molecule, using the symbols in the examples above. Some motions may be out-of-plane motions, so be sure to rotate the molecules on the computer screen to examine them from different angles.

- Record the frequency for each vibration. Many programs give the frequencies in cm^{-1} (wavenumber); this is actually the frequency, in s^{-1}, divided by the speed of light (3.00×10^{10} cm/s). For this requirement, just write down whatever the program says; if necessary, you can convert units later.

Requirement 3: Symmetry Elements and Symmetry Operations

Because the molecules that we are examining are very small, the rules of quantum mechanics govern the processes in which we are interested. According to quantum mechanics, not all light absorption processes are allowed; many are "forbidden" by symmetry. To determine which molecular vibrations absorb infrared light, we need to examine the actions of the symmetry operations of the molecules on the coordinate axes and on the molecular vibrations. We begin by determining the symmetry elements and operations of each molecule.

A *symmetry operation* on a molecule is an action that moves the molecule into a position indistinguishable from the starting position. A *symmetry element* of a molecule is a geometric feature of the molecule about which a symmetry operation is performed. Symmetry elements include planes and axes; symmetry operations include reflections across planes and rotations about axes. In the case of water, 180° rotation about the z-axis is a symmetry operation, denoted $\hat{C}_2$, while the z-axis itself is a symmetry element, C_2. The symbol C_3 indicates a three-fold axis of symmetry, a symmetry element, while $\hat{C}_3$ indicates a 120° rotation about C_3, a symmetry operation. Rotation by 240° about C_3 is denoted $\hat{C}_3^2$. Rotation by 360° about C_3 is equivalent to doing nothing; that is, $\hat{C}_3^3 = \hat{E}$, where $\hat{E}$ is the identity operation.

Table 1.
A list of all symmetry elements and operations.

Symmetry Elements}		Symmetry Operations	
Symbol	Description	Symbol	Description
E	Identity	$\hat{E}$	No change
C_n	n-fold axis of symmetry	$\hat{C}_n$	Rotation about the axis by 360°/n
σ	Plane of symmetry	$\hat{\sigma}$	Reflection through the plane
i	Center of symmetry	$\hat{i}$	Reflection through the center
S_n	n-fold rotation-reflection axis of symmetry, also called an *improper* rotation axis	$\hat{S}_n$	Rotation about the axis by 360°/n followed by a reflection through a plane perpendicular to that axis

The plane of symmetry, σ, is also referred to as a *reflection plane* or *mirror plane*. The symbol σ_v is used to denote a "vertical" plane of symmetry that is parallel to an axis of highest symmetry (z-axis, or C_n with largest n), while the symbol σ_h is used to denote a "horizontal" plane of symmetry that is perpendicular to an axis of highest symmetry. The symbol σ_d denotes a "dihedral" plane of symmetry that bisects an angle between atoms.

Example 3: Figure 6 shows the symmetry elements and operations of water.

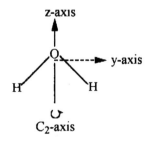

Figure 6. Symmetry elements of water.

The symmetry elements are E, C_2 (shown), σ_v (xz-plane, perpendicular to the paper), and $\sigma_{v'}$ (yz-plane, the plane of the paper). Note that it does not matter whether σ_v represents the xz- or the yz-plane. The corresponding symmetry operations for water are $\hat{E}$, $\hat{C}_2$, $\hat{\sigma}_v$, and $\hat{\sigma}_{v'}$.

The symmetry elements not shown are E, σ_v (yz-plane, the plane of the paper), σ_h (xy-plane, perpendicular to the C_∞ axis), and i (center of inversion at the coordinate origin).

Example 4: Figure 7 shows the symmetry elements and operations of carbon dioxide.

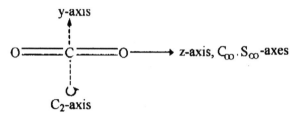

Figure 7. Symmetry elements of carbon dioxide.

The symmetry operations are $\hat{E}$, infinitely many $\hat{C}_2$, $\hat{C}_\infty$, infinitely many $\hat{\sigma}_v$, $\hat{S}_\infty$, $\hat{\sigma}_h$, and $\hat{\imath}$. A subscript of ∞ means that rotation through any angle about that axis results in a valid symmetry operation. Note that $\hat{\sigma}_h$ is identical to $\hat{S}_0$ and is often omitted.

Re-examine the ball-and-stick models for your molecules. **Determine all of the symmetry elements and corresponding symmetry operations for each molecule.** (Hint: Two molecules on the list contain the S_4 symmetry element and one contains an S_6 symmetry element.)

There are a few things to keep in mind while trying to determine the symmetry elements for chemical compounds:

- Molecules are three-dimensional objects. This means that we can tell the difference between the "front" and "back" or the "top" and "bottom" of planar molecules. For instance, the $\hat{\sigma}_{v'}$ reflection of the water molecule across the yz-plane is not the same as the identity operation $\hat{E}$.

- Since atoms of the same kind (or color) are indistinguishable, you may want to number the atoms in the models in order to keep track of the results of the symmetry operations.

- When molecules have hexagonal rings with alternating double bonds, all of these bonds—both single and double—are equivalent (e.g., benzene and toluene). It is only the orientation of the atoms themselves that can be "seen" spectroscopically and hence needs to be considered here.

After you have determined the symmetries of the molecules, you can double-check your axis assignments. The z-axis should be the axis of highest C_n symmetry. In H_2O, there is only one C_n axis, C_2, so it is the z-axis. In CO_2, there is a C_2 axis and a C_∞ axis, so the C_∞ axis is the z-axis. Check the other molecules to make sure that the z-axis that you assigned is the one of highest symmetry. Remember that the axis of highest symmetry may not be unique.

Requirement 4: Orders of Symmetry Operations

The *order* of a symmetry operation is the number of times that the operation must be applied to obtain the identity operation, $\hat{E}$. More specifically, the order of a symmetry operation $\hat{A}$ is the smallest positive integer n such that $\hat{A}^n = \hat{E}$. For instance, for H_2O the nonidentity symmetry operations each have order 2. Note that an inversion always has order 2. The symmetry operation $\hat{S}_4$ has order 4 because it must be applied four times in succession to return the molecule to its original orientation when the outside atoms are labeled. (Try it for allene or methane!) Therefore, $\hat{S}_4$ generates 4 symmetry operations: $\hat{S}_4$, $\hat{S}_4^2 = \hat{C}_2$, $\hat{S}_4^3$, and $\hat{S}_4^4 = \hat{E}$ of orders 4, 2, 4, and 1, respectively. **For your molecules, find the order of each symmetry operation.**

Requirement 5: Symmetry Groups

You may have noticed that the set of symmetry operations forms a group under composition of operations, called the *symmetry group* of the molecule. Consider the groups of symmetries of the molecules water, methyl bromide, dichlorodifluoromethane, ethylene, allene, sulfur trioxide, and toluene (sharing data with other students if necessary). What is the order of each group? Which groups are abelian? Use all of the information you've accumulated so far (order of group, orders of its elements, whether or not it's abelian) to identify each group with the name you gave it in Abstract Algebra class. For instance, the symmetry group for H_2O is a group of order 4 in which each nonidentity element has order 2. Up to isomorphism, there are just two groups of order 4: the cyclic group, Z_4, and the Klein group, V_4. **Which is the symmetry group for water? Now determine the symmetry groups for the other six chemicals listed above.**

Requirement 6: Action of Symmetries on Coordinate Axes

Your overall goal is to identify the molecular origins of the infrared peaks that you observe in air spectra. Since a molecule's infrared light absorption depends on how the Cartesian axes transform under the symmetry operations, the next step is to determine what happens to each of the Cartesian axes as the different symmetry operations are performed upon the molecule.

Example 5: In **Figure 8** and in **Table 2**, we illustrate this process for water. Note that under the identity operation, $\hat{E}$, none of the axes is inverted or reversed. When the molecule is rotated about the C_2 axis, the orientation of the z-axis remains the same but the x- and y-axes are oriented in the opposite directions; each point (x, y, z) in R^3 is moved to the point $(-x, -y, z)$. In this case, the rotation is equivalent to multiplying the x and y values by -1. For any operation, -1 indicates that there is a reversal in the orientation of the axis relative to the original orientation, whereas $+1$ indicates that the orientation remains the same.

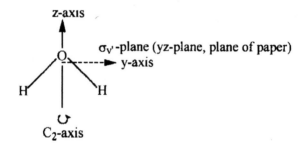

Figure 8. The coordinate axes and symmetry elements for water.

Table 2.

How the axes of water transform under the symmetry operations.

H_2O	$\hat{E}$	$\hat{C}_2$	$\hat{\sigma}_v \, (xz)$	$\hat{\sigma}_{v'} \, (yz)$
x	1	-1	1	-1
y	1	-1	-1	1
z	1	1	1	1

Construct similar tables for toluene, dichlorodifluoromethane, and ethylene. First check with your instructor to make sure you have drawn the Cartesian coordinate axes in the standard way for each of these molecules.

Requirement 7: Action of Symmetries on Vibrational Motions

The vibrational motions of the molecule that absorb infrared radiation are the ones that transform under the symmetry operations of the molecule in the same way as do the Cartesian coordinate axes of the molecule. Therefore, we need to determine how the molecular vibrations behave under each of the symmetry operations, so that we can compare them to the transformations of the Cartesian coordinate axes.

Example 6: Let us re-examine the vibrations of water. We reproduce **Figure 4** for convenience:

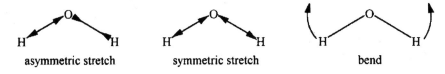

asymmetric stretch	**symmetric stretch**	**bend**

Figure 4. Vibrations of water.

If we ask how each of the vibrations of water behaves under each of the symmetry operations, we can add more entries to **Table 2**. When we examine what happens to the vibrating molecules as the symmetry operations are performed, we are interested only in whether or not the geometrical orientation of the molecule has changed. For instance, consider the water molecule as it undergoes the asymmetric stretch. Imagine the molecule (or stop the computer program) when it reaches its most extreme position, with one "arm" extended and the other "arm" contracted. Now perform the $\hat{C}_2$ operation (rotation by 180°) on this "distorted" molecule. Its orientation after the $\hat{C}_2$ operation is different from its orientation before. Note that the configuration has been reversed. The fact that its new position is distinguishable from its original position is represented in **Table 3** by -1.

Now let's examine the water molecule as it undergoes the bend or symmetric stretch. If we perform the $\hat{C}_2$ operation (rotation by 180°) on this vibrating molecule, its geometric orientation is unchanged. Its new position is indistinguishable from its original position. The fact that it appears unchanged is represented in **Table 3** by +1.

Table 3 is for water; construct tables like it for toluene, dichlorodifluoromethane, and ethylene. Examine the motions of the atoms for each different vibration. If you were to imagine the molecule (or stop the computer program) when it reaches its most extreme position, consider how that "version" of the molecular shape would behave under each of the different symmetry operations. That the molecule is indistinguishable after the symmetry operation is

Table 3.

How the axes and vibrations of water transform under the symmetry operations.

H_2O	$\hat{E}$	$\hat{C}_2$	$\hat{\sigma}_v\,(xz)$	$\hat{\sigma}_{v'}\,(yz)$
x	1	−1	1	−1
y	1	−1	−1	1
z	1	1	1	1
bend	1	1	1	1
symmetric stretch	1	1	1	1
asymmetric stretch	1	−1	−1	1

indicated by a 1, while a distinguishable molecule is represented by a −1. Fill in a line in your table for each vibrational motion.

Requirement 8: Comparing Symmetry Operations on Axes and Vibrations

For the molecules toluene, dichlorodifluoromethane, and ethylene, use the tables that you constructed in **Requirement 7** to determine which of the vibrational motions transform under the symmetry operations as do the x-, y-, or z-axes, and, for those that do, list the frequency calculated for each motion. (You listed the frequencies you need for this problem in **Requirement 2**.) These are the frequencies that absorb light. Given that infrared spectrometers operate in the range of about 600 cm^{-1} to 4000 cm^{-1} (wavenumber), which vibrational frequencies should you observe in the infrared spectra for your molecules? **Make a rough sketch of the infrared spectrum that you would expect to see for each of your molecules, labeling peaks with their frequencies.** Assume that if peaks are separated by less than 25 cm^{-1}, they will not be resolved and will appear as a single peak.

> **Example 7:** In the case of water, as illustrated in **Table 3**, the bending vibration transforms under the symmetry operations in the same way as does the z-axis. This is also true for the symmetric stretching motion. On the other hand, the asymmetric stretch transforms in the same way as does the y-axis. In this case, all three of the vibrational motions of water would absorb infrared light since each one of them transforms under the symmetry operations as does one of the Cartesian coordinate axes. All of the vibrations also have frequencies that are in the appropriate frequency range. We would expect the infrared spectrum of water to have three peaks corresponding to frequencies of 1840, 3587, and 3652 cm^{-1}.

Requirement 9: Actions of Symmetries on the xy-plane

Now look at CH_3Br, methyl bromide. Notice that the orientation of the z-axis does not change under any of the symmetry operations of this molecule; therefore, we need to worry only about transformations of the xy plane, or R^2. For this molecule, some of the symmetry operations move the x- and y-axes to positions that are not 180° from their original orientations. In this case, we describe all movements of the xy-plane using 2×2 transformation matrices. For example, the operation $\hat{C}_3$ rotates the molecule by an angle of $\theta = 120°$ in the counterclockwise direction, about its z-axis. Hence, the x- and y-axes are rotated as shown in **Figure 9**.

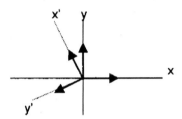

Figure 9. Rotation of the x- and y-axes through 120°.

Since vectors $\mathbf{i} = (1, 0)$ and $\mathbf{j} = (0, 1)$ form a basis for R^2, linear transformations of R^2 are completely determined by where they map the basis vectors $\mathbf{i}$ and $\mathbf{j}$. The 120° rotation maps vector $\mathbf{i}$ to the vector with x-component $\cos 120°$ $(= -\sin 30°)$ and y-component $\sin 120°$ $(= \cos 30°)$, that is, to the vector $(-1/2, \sqrt{3}/2)$. Similarly, vector $\mathbf{j}$ is mapped to the vector with x-component $\cos 210°$ $(= -\cos 30° = -\sin 120°)$ and y-component $\sin 210°$ $(= -\sin 30° = \cos 120°)$, that is, to the vector $(-\sqrt{3}/2, -1/2)$. Considering the points in the xy-plane as column vectors, rotation by $\theta = 120°$ is given by left multiplication of each point by the matrix

$$\begin{bmatrix} -\dfrac{1}{2} & -\dfrac{\sqrt{3}}{2} \\ \dfrac{\sqrt{3}}{2} & -\dfrac{1}{2} \end{bmatrix}.$$

That is,

$$\begin{bmatrix} -\dfrac{1}{2} & -\dfrac{\sqrt{3}}{2} \\ \dfrac{\sqrt{3}}{2} & -\dfrac{1}{2} \end{bmatrix} \begin{bmatrix} 1 \\ 0 \end{bmatrix} = \begin{bmatrix} -\dfrac{1}{2} \\ \dfrac{\sqrt{3}}{2} \end{bmatrix}, \qquad \begin{bmatrix} -\dfrac{1}{2} & -\dfrac{\sqrt{3}}{2} \\ \dfrac{\sqrt{3}}{2} & -\dfrac{1}{2} \end{bmatrix} \begin{bmatrix} 0 \\ 1 \end{bmatrix} = \begin{bmatrix} -\dfrac{\sqrt{3}}{2} \\ -\dfrac{1}{2} \end{bmatrix},$$

and

$$\begin{bmatrix} -\dfrac{1}{2} & -\dfrac{\sqrt{3}}{2} \\ \dfrac{\sqrt{3}}{2} & -\dfrac{1}{2} \end{bmatrix} \begin{bmatrix} a \\ b \end{bmatrix} = \begin{bmatrix} -\dfrac{1}{2}a - \dfrac{\sqrt{3}}{2}b \\ \dfrac{\sqrt{3}}{2}a - \dfrac{1}{2}b \end{bmatrix}.$$

In general, rotation of the xy-plane about the origin by an angle θ in the counterclockwise direction can be described by the matrix

$$\begin{bmatrix} \cos\theta & -\sin\theta \\ \sin\theta & \cos\theta \end{bmatrix}.$$

To determine matrices for reflections across planes of symmetry, we first reorient the methyl bromide molecule so that one of the hydrogen atoms lies along the positive x-axis, as shown in **Figures 10** and **11**. Note that **Figure 11** shows the methyl bromide molecule projected onto the xy-plane.

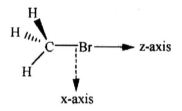

methyl bromide

Figure 10. Reoriented methyl bromide.

As seen in **Figure 11**, methyl bromide now has hydrogen atoms lying along the positive x-axis, at an angle of 120° counterclockwise from the positive x-axis (along the dotted line), and at an angle of 240° counterclockwise from the positive x-axis.

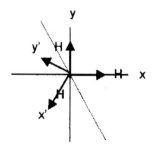

Figure 11. Reflection transformation for CH_3Br, methyl bromide.

Methyl bromide has vertical planes of reflection along each H-C-Br angle as symmetry elements. For a reflection such as the one across the vertical symmetry plane through the dotted line, we must determine where the vectors $\mathbf{i} = (1, 0)$ and $\mathbf{j} = (0, 1)$ are mapped.

Vector $\mathbf{i}$ is mapped by the transformation to the vector $(-1/2, -\sqrt{3}/2)$. Similarly, vector $\mathbf{j}$ is mapped to the vector $(-\sqrt{3}/2, 1/2)$. The transformation matrix is

$$
\begin{bmatrix}
-\dfrac{1}{2} & -\dfrac{\sqrt{3}}{2} \\[2ex]
-\dfrac{\sqrt{3}}{2} & \dfrac{1}{2}
\end{bmatrix}.
$$

Table 4 shows how the coordinate axes of methyl bromide transform under the symmetry operations of the molecule. **Find the three remaining** 2×2 **matrices, using the method illustrated above.**

Table 4.
Axis-transformation table for CH_3Br (matrices).

CH_3Br	$\hat{E}$	$\hat{C}_3$	$\hat{C}_3^2$	$\hat{\sigma}_v$	$\hat{\sigma}_{v'}$	$\hat{\sigma}_{v''}$
z	(1)	(1)	(1)	(1)	(1)	(1)
x, y	$\begin{bmatrix} 1 & 0 \\ 0 & 1 \end{bmatrix}$	$\begin{bmatrix} -\frac{1}{2} & -\frac{\sqrt{3}}{2} \\ \frac{\sqrt{3}}{2} & -\frac{1}{2} \end{bmatrix}$				$\begin{bmatrix} -\frac{1}{2} & -\frac{\sqrt{3}}{2} \\ -\frac{\sqrt{3}}{2} & \frac{1}{2} \end{bmatrix}$

Table 5 is an axis-transformation table for methyl bromide in which the matrices are replaced by their traces. **Fill in the missing entries in Table 5.**

Table 5.
Axis-transformation table for CH_3Br (traces).

CH_3Br	$\hat{E}$	$\hat{C}_3$	$\hat{C}_3^2$	$\hat{\sigma}_v$	$\hat{\sigma}_{v'}$	$\hat{\sigma}_{v''}$
z	1	1	1	1	1	1
x, y	2	-1				0

Now that we know how the co-ordinate axes transform under the symmetry operation of CH_3Br, we must determine which vibrational motions transform in the same way as the axes. Re-examine the vibrational motions for methyl bromide, CH_3Br, whose frequencies are unique (i.e., do not occur in pairs). Which of the vibrations transform in the same way as does the z-axis?

For this molecule, which is three-fold symmetric, it is more difficult to try to determine which motions transform in the same way as do the x- and y-axes. Using the molecular modeling program, orient the methyl bromide molecule so that you are looking at it with the hydrogens in front and the bromine in the back—in other words, you are viewing the xy-plane with the negative z-axis pointing towards you. Observe the six motions whose frequencies occur in pairs; these motions are called *degenerate* motions.

If you consider the two motions in each degenerate pair as a set, you will notice that the motions that occur during the vibrations are along either the x-axis or the y-axis. If you examine the lowest frequency pair, you should notice that the vibrational motions involve H-C-Br angle changes ($\sim$850 cm^{-1}). If you orient the molecule so that this motion occurs primarily in the x-axis direction and then (without moving the molecule) switch to the second motion of the pair, you will notice that the motion is now along the y-axis. Try this with the pair of next highest frequency (the H-C-H bending motion, $\sim$1350 cm^{-1}). You should notice the same thing. Now examine the highest frequency pair (H-C stretch, $\sim$3150 cm^{-1}). Does the same thing occur?

The atoms within a molecule are held together by their electrons (which have negative charges); the electrons form the chemical bonds. As the atoms in the molecule move during a vibration, so do the electrons. It is this oscillating electron density that interacts orthogonally with the oscillating electromagnetic field of the light. For molecules such as methyl bromide, in which the x- and y-axes move together under the symmetry operations, we must have electron density oscillation in both the x- and y-directions in order to allow light absorption. In the case of methyl bromide, one member of each degenerate pair has electron density oscillation in the x-direction while the other has oscillation in the y-direction. Therefore, these vibrations transform as do the x- and y-axes.

List all the frequencies that you should see in the infrared spectrum of CH$_3$Br, then sketch the spectrum that you would expect to see, labeling the peaks with their corresponding frequencies.

For methyl bromide, all of the degenerate vibrations transform in the same way as do the scrambled x- and y-axes. This is not always true, however. Examine the degenerate vibrations for benzene. If you examine the degenerate pairs of vibrations whose frequencies are approximately 1100 cm^{-1}, 1650 cm^{-1}, and 3240 cm^{-1} (and are labeled E_{1u}), you should notice that they all have distinct electron density oscillations in the x- and y-directions. Now examine the other 7 pairs of degenerate vibrations. Can you see that they do not have oscillating electron densities along each of the x- and y-axes?

Requirement 10: Obtaining Infrared Spectra

Now it is time to compare your predicted spectra to real ones. Most of the chemicals that you have studied are common and available in chemistry departments, as are infrared absorption spectrometers. **Obtain the spectra of as many of your molecules as you can on the chemistry department's infrared spectrometer. Do you observe absorptions where you had predicted?** If not, try multiplying all of your calculated vibrational frequencies by 0.89 (a factor that theoretical chemists recommend to compensate for overcalculations). Do these frequencies come closer to what you observe?

Requirement 11: Identifying the Pollutant

Examine the spectrum in **Figure 12**, from your sampling trip to detect atmospheric pollutants. Note that it is a very different spectrum from that of carbon dioxide and water in **Figure 1**, which we reproduce again for easy comparison. In **Figure 12**, the spectrometer has automatically electronically subtracted the "background" spectrum of air, which is mostly carbon dioxide and water.

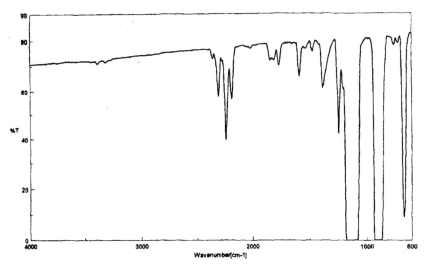

Figure 12. Infrared spectrum of an atmospheric pollutant.

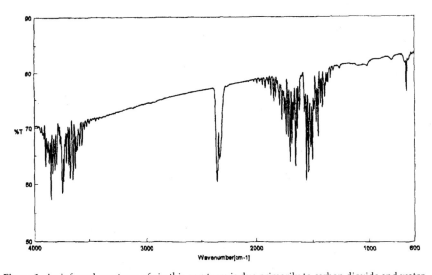

Figure 1. An infrared spectrum of air; this spectrum is due primarily to carbon dioxide and water.

When you took the spectrum in **Figure 12**, you set up the mirror 20 m away across an open farm field where strawberries were growing. In the field on one side, there were cows grazing; on the other side, there was a natural-gas pumping station. Across the street, there was a gas station with an auto repair shop specializing in air-conditioner repair. This information narrows the list of suspected chemicals to six. **Which ones remain on the list?** (Remember, the carbon dioxide has been subtracted from the spectrum already.) **Which of the chemicals that you studied is responsible for this spectrum? To which potential polluter would you attribute this pollution?**

Reference

McQuarrie, Donald A., and John D. Simon. 1997. *Physical Chemistry: A Molecular Approach.* Sausalito, CA: University Science Books.

Acknowledgments

We would like to thank Dr. Christopher Brazier for his careful reading of this manuscript and for his helpful suggestions. We also would like to thank Dr. Steven Morics and Dr. Teresa Longin and their students for testing early versions of this project.

Title: Pollution Police

Sample Solution

Requirement 1: Diagrams of the nine remaining molecules with their Cartesian coordinate axes are shown in **Figure S1**. Note that the positive direction chosen for each axis is for convenience of illustration only; should a student choose a different orientation for any axis, it would make no difference in the frequency analysis. If a student interchanges any axes, this changes only the Cartesian coordinate markers indicated in **Requirements 6, 7, and 8**; before students get to those problems, you might check to make sure that they have the axes labeled in the standard way, to make grading easier.

Methane: All C-H bonds in methane are equivalent, so the z-axis may be placed along any C-H bond. However, in our two-dimensional representation, we have chosen to use one of the bonds lying in the plane of the paper for convenience. In the methane molecule, the x-axis points out of the plane of the paper toward you, as does any axis not shown in each of the remaining molecules.

Sulfur hexafluoride: All axes in SF_6 are equivalent, so the z-axis could be placed along any of the three F-S-F linear bonds. Our rules for placing axes require that the x- and y-axes also be placed along F-S-F bonds.

Methyl bromide: Students probably will draw the axes for methyl bromide as shown at left in **Figure S1** (perhaps with the y-axis pointing upward). This is consistent with the rules given in the problem statement; however, in **Requirement 9**, the x- and y-axes will be interchanged, as shown at right, for ease of computation.

Allene: Although students probably will draw the axes of allene as shown in **Figure S1**, the convention is to place the x- and y-axes so that they coincide with the C_2 rotational axes for the molecule. This requires that the x- and y-axes be placed along dihedral angles, that is, that they be "tilted" 45° forward (or backward) from what is indicated here.

Requirement 2: This problem requires the use of a molecular modeling program available in many chemistry departments. Consult your chemistry colleagues for help with this program. Molecular modeling programs offer many different levels of calculational sophistication. In the interest of time, we recommend that you use the lowest level available in the program to perform these calculations. Since it takes quite a bit of time to diagram all of the vibrations for each molecule, you may wish to have students record vibrational motions and frequencies only for water, dichlorodifluoromethane, ethylene, methyl bromide, and toluene. (Note that while dichlorodifluoromethane and methyl bromide have 9 vibrational motions and ethylene has 12 vibrational

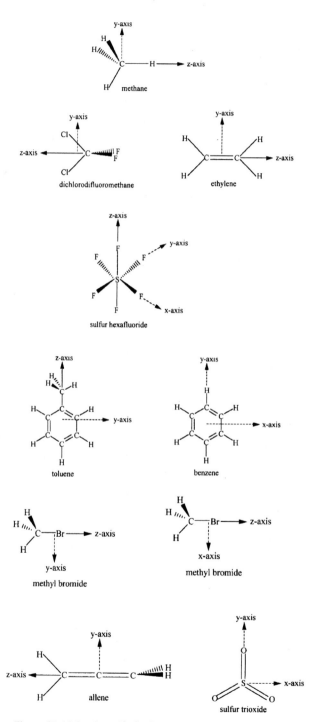

Figure S1. Molecules with the Cartesian co-ordinates included.

motions, toluene has 39 different vibrational motions!) The students do not need to record the motions of the molecules for use later if they have ongoing access to the molecular modeling program. That is, they may return to the program to complete subsequent problems without physically recording the motions ahead of time.

Different programs may output the vibrational frequencies in different units. Wavenumber (cm^{-1}) is actually the frequency (in s^{-1}) divided by the speed of light (3.00×10^{10} cm/s). If the wavelength is given, you can convert it into cm (probably from μm) and then take its reciprocal to obtain wavenumber.

Requirement 3: Using the ball-and-stick molecular models is very helpful for this problem, as is numbering identical atoms. To speed up student work, you may wish to tell them ahead of time how many symmetry operations each molecule has. Since we will not ask students to work further with molecules with symmetry groups of order larger than 12, you may want to instruct students not to spend too much time working with these molecules.

The set of all symmetry operations of a molecule forms a group called the *symmetry group* or *point group* of the molecule. For example, the symmetry group for water is denoted by chemists as C_{2v} and by mathematicians as V_4 or $Z_2 \oplus Z_2$. These chemical symmetries are not the same as the symmetry elements and operations; in fact, they correspond to the irreducible characters of the group. The numbers listed in the table are the values of the irreducible characters on each conjugacy class. It turns out that the normal modes of vibration of a molecule correspond to irreducible representations of the symmetry group to which the molecule belongs. For more information, consult the **Further Reading** list.

Water, toluene, and dichlorodifluoromethane. The symmetry groups have order 4. They belong to the group C_{2v}; have symmetry elements E, C_2 (z-axis), σ_v (xz-plane), and $\sigma_{v'}$ (yz-plane); and have symmetry operations $\hat{E}$, $\hat{C}_2$, $\hat{\sigma}_v$, and $\hat{\sigma}_{v'}$. (It does not matter whether σ_v represents the xz- or the yz-plane.)

Methyl bromide (or bromo-methane). The symmetry group has order 6. Methyl bromide belongs to the C_{3v} group; has symmetry elements $\hat{E}$, $\hat{C}_3$, and 3 different σ_v planes (each containing a different Br-C-H angle); and has symmetry operations $\hat{E}$, $\hat{C}_3$, $\hat{C}_3^2$, $\hat{\sigma}_v$, $\hat{\sigma}_{v'}$, and $\hat{\sigma}_{v''}$. In chemical character tables (e.g., Cotton [1990]), the symmetry operations $\hat{C}_3$ and $\hat{C}_3^2$ are listed as $2C_3$.

Ethylene (or ethene). The symmetry group has order 8. Ethylene belongs to the D_{2h} group; has symmetry elements E, 3 different C_2 axes (each along a Cartesian coordinate axis), i, and 3 different σ_v planes (xy, xz, and yz); and has the symmetry operations $\hat{E}$, $\hat{C}_2$, $\hat{C}_{2'}$, $\hat{C}_{2''}$, $\hat{i}$, $\hat{\sigma}_v$, $\hat{\sigma}_{v'}$, and $\hat{\sigma}_{v''}$.

Allene (or 1,2-propadiene). The symmetry group has order 8. Allene belongs to the D_{2d} group and has symmetry elements E, an S_4 axis, 3 C_2 axes along the

Cartesian coordinate axes, and 2 σ_d planes (the two planes defined by the two different CH_2 groups at the ends of the molecule). Remember that the x- and y-axes are dihedral; they bisect the $90°$ angles between the two planes defined by the two different CH_2 groups at the ends of the molecule. The symmetry operations are $\hat{E}$, $\hat{S}_4$, $\hat{S}_4^2 = \hat{C}_2$, $\hat{S}_4^3$, $\hat{C}_{2'}$, $\hat{C}_{2''}$, $\hat{\sigma}_d$, and $\hat{\sigma}_d'$.

Sulfur trioxide. The symmetry group has order 12. Sulfur trioxide belongs to the D_{3h} group and has symmetry elements E, C_3, 3 different C_2 axes (along each S-O bond), σ_h (the plane of the molecule), S_3, and 3 different σ_v planes (each containing a different C_2 axis). It has symmetry operations $\hat{E}$, $\hat{C}_3$, $\hat{C}_3^2$, $\hat{C}_2$, $\hat{C}_{2'}$, $\hat{C}_{2''}$, $\hat{\sigma}_h$, $\hat{S}_3$, $\hat{S}_3^5 = \hat{S}_3^{-1}$ (rotation by $-120°$ followed by a single reflection), $\hat{\sigma}_v$, $\hat{\sigma}_{v'}$, and $\hat{\sigma}_{v''}$.

Methane. The symmetry group has order 24. Methane belongs to the T_d group and has symmetry elements E, 4 different C_3 axes (along each C-H bond), 3 different C_2 axes (all of which are located in dihedral orientations, or bisecting the H-C-H angles), 3 different S_4 axes (aligned along the C_2 axes), and 6 σ_d planes (which lie along the planes formed by the H-C-H angles). It has symmetry operations $\hat{E}$, 4 different $\hat{C}_3$, 4 different $\hat{C}_3^2$, $\hat{C}_2$, $\hat{C}_{2'}$, $\hat{C}_{2''}$, 3 different $\hat{S}_4$, 3 different $\hat{S}_4^3$, and 6 different $\hat{\sigma}_d$.

Benzene. The symmetry group has order 24. Benzene belongs to the D_{6h} group and has symmetry elements E, C_6 (z-axis, perpendicular to the plane of the paper), C_3 (coincident with C_6), C_2 (coincident with C_6), 3 different $C_{2'}$ axes (through opposite H-C bonds), 3 $C_{2d} = C_{2''}$ axes (bisecting C-C bonds), i, S_6 (coincident with the C_6 axis), S_3 (coincident with the C_6 axis), σ_h (perpendicular to the C_6 axis), 3 different σ_v planes (containing opposite pairs of C-H bonds), and 3 σ_d planes (bisecting opposite C-C bonds, hence the d designation since they are dihedral planes). It has symmetry operations $\hat{E}$, $\hat{C}_6$, $\hat{C}_6^2 = \hat{C}_3$, $\hat{C}_6^3 = \hat{C}_2$, $\hat{C}_6^4 = \hat{C}_3^2$, $\hat{C}_6^5$, 3 different $\hat{C}_{2'}$, 3 different $\hat{C}_{2d}$, $\hat{i}$, $\hat{S}_6$, $\hat{S}_6^5$, $\hat{S}_3$, $\hat{S}_3^5$, $\hat{\sigma}_h = \hat{S}_6^3$, $\hat{\sigma}_v$, $\hat{\sigma}_{v'}$, $\hat{\sigma}_{v''}$, $\hat{\sigma}_d$, $\hat{\sigma}_{d'}$, and $\hat{\sigma}_{d''}$.

Sulfur hexafluoride. The symmetry group has order 48. Sulfur hexafluoride belongs to the O_h group and has symmetry elements E, 3 different C_4 axes (along the F-S-F bonds), 3 different C_2 axes (each coincident with a C_4), 4 different C_3 axes (perpendicular to the faces of the octahedron—to see these, set the model of the molecule on a desk and look down!), 6 different $C_{2d} = C_{2'}$ axes (perpendicular to the edges of the octahedron), i, 4 different S_6 axes (coincident with the C_3 axes), 3 different S_4 axes (coincident with the C_4 axes), 3 different σ_h planes (perpendicular to each C_4), and 6 different σ_d planes (containing each C_4 axis and bisecting the F-S-F angles—in other words, bisecting the edges of the octahedron). It has symmetry operations $\hat{E}$, 3 $\hat{C}_4$, 3 $\hat{C}_4^3$, 3 $\hat{C}_4^2 = \hat{C}_2$, 4 $\hat{C}_3$, 4 $\hat{C}_3^2$, 6 different $\hat{C}_{2d}$, $\hat{i}$, 4 different $\hat{S}_6$, 4 different $\hat{S}_6^5$, $\hat{S}_4$, $\hat{S}_4'$, $\hat{S}_4''$, $\hat{S}_4^3$, $\hat{S}_4^{3'}$, $\hat{S}_4^{3''}$, $\hat{\sigma}_h$, $\hat{\sigma}_h'$, $\hat{\sigma}_{h''}$, and 6 different $\hat{\sigma}_d$. Note that $\hat{S}_6^2 = \hat{C}_3$, $\hat{S}_6^3 = \hat{i}$, and $\hat{S}_6^4 = \hat{C}_3^2$.

Requirement 4:

$\hat{E}$: order 1

$\hat{\sigma}, \hat{i}, \hat{S}_2, \hat{C}_2$: order 2

$\hat{C}_3, \hat{C}_3^2$: order 3

$\hat{S}_4, \hat{S}_4^3, \hat{C}_4, \hat{C}_4^3$: order 4

$\hat{S}_4^2, \hat{C}_4^2$: order 2

$\hat{S}_6, \hat{S}_6^5, \hat{C}_6, \hat{C}_6^5$: order 6

$\hat{S}_6^2, \hat{S}_6^4, \hat{C}_6^2, \hat{S}_6^4$: order 3

$\hat{S}_6^3, \hat{C}_6^3$: order 2

$\hat{S}_3, \hat{S}_3^5$: order 6

$\hat{S}_3^2, \hat{S}_3^4$: order 3

$\hat{S}_3^3$: order 2

Requirement 5: The order of the group is the number of different symmetry operations for that group. Group orders were given in the solution to **Requirement 3** and appear in **Table S1**.

Table S1.
Solution to **Requirement 5**.

Molecule	Abelian?	Group	Additional Evidence
Water	yes	Klein group V_4 or $Z_2 \oplus Z_2$	Each nonidentity element has order 2, group has order 4
Methyl bromide	no	Symmetric group on 3 letters S_3 or dihedral group of order 6, D_3	Group has order 6; there is only one nonabelian group of order 6
Dichlorodifluoro-methane	yes	V_4 or $Z_2 \oplus Z_2$	Each nonidentity element has order 2, group has order 4
Ethylene	yes	$Z_2 \oplus Z_2 \oplus Z_2 \oplus Z_2$	Each nonidentity element has order 2, group has order 8
Allene	no	Dihedral group of order 8, D_4	Group has 2 elements of order 4 and 5 of order 2, group has order 8; the other nonabelian group of order 8 has more than 2 elements of order 4
Sulfur trioxide	no	Dihedral group of order 12, D_6, or $D_3 \oplus Z_2$	Group has 2 elements of order 6, 2 of order 3, and 7 of order 2; the other nonabelian groups of order 12 have elements of order 4
Toluene	yes	V_4 or $Z_2 \oplus Z_2$	Each nonidentity element has order 2, group has order 4

Requirement 6: Both toluene and dichlorodifluoromethane belong to the same group as water, so the axes transform in the same way; see **Table S2**.

What students are constructing here is part of the chemical character table of each molecule. Note that the x, y, and z rows of **Table S2** correspond to the rows marked x, y, and z in the C_{2v} chemical character table. Infrared spectroscopy requires a single photon of light and hence is able to excite any vibration that transforms as does at least one of the Cartesian coordinate axes. For ethylene, the axes transform in the way shown in **Table S3**.

Table S2.

How the axes transform for toluene and dichlorodifluoromethane.

toluene and dichlorodifluoromethane	$\hat{E}$	$\hat{C}_2$	$\hat{\sigma}_{v'}\ (xz)$	$\hat{\sigma}_v\ (yz)$
x	1	-1	1	-1
y	1	-1	-1	1
z	1	1	1	1

Table S3.

How the axes transform for ethylene.

C_2H_4	$\hat{E}$	$\hat{C}_2\ (z)$	$\hat{C}_2\ (y)$	$\hat{C}_2\ (x)$	$\hat{\imath}$	$\hat{\sigma}\ (xy)$	$\hat{\sigma}\ (xz)$	$\hat{\sigma}\ (yz)$
x	1	-1	-1	1	-1	1	1	-1
y	1	-1	1	-1	-1	1	-1	1
z	1	1	-1	-1	-1	-1	1	1

Requirement 7: See **Tables S4–S6.** We have numbered the vibrational modes from lowest to highest calculated frequency; these numbers appear in the left-hand columns of **Tables 4–6.** Remember that the calculated frequencies and the order in which they are listed may be different in different programs. The frequency values found in the spectra should be 80% to 95% of those calculated by the molecular modeling programs; this is because the quantum mechanical calculations used in these programs overestimate all molecular energies.

Requirement 8: See **Tables S7–S9.** All listed frequencies are in cm^{-1}, wavenumber.

Toluene: The spectrum should have peaks at 709.1, 829.8 (average of 826.4 and 833.2), 900.7, 1088.3 (average of 1071.7, 1094.6, and 1098.5), 1136.4 (average of 1126.4 and 1146.4), 1210.9, 1238.1 (average of 1236.4 and 1239.7), 1274.9, 1321.3, 1363.2, 1507.5, 1591.2, 1608.5, 1667.6 (average of 1656.6, 1669.2, and 1677.0), 1721.6 (average of 1715.6 and 1727.5), 3015.9, 3075.5 (average of 3068.0 and 3082.9), and 3242.0 (average of 3225.4, 3233.4, 3251.8, and 3257.4) cm^{-1}. The first six frequencies are too low to be detected by the spectrometer.

CCl$_2$F$_2$: The spectrum should have peaks at 689.9 cm^{-1}, 868.6 cm^{-1}, 1388.1 cm^{-1}, and 1505.5 cm^{-1}; the other peaks are out of the range of the spectrometer.

C$_2$H$_4$: You would expect peaks at 823.6, 1028.9, 1385.0, and 3256.8 cm^{-1}.

Students should provide a sketch of their predicted spectrum for each of the three molecules with downward (or upward) peaks corresponding to each of the frequencies listed above.

Table S4.

Toluene.

toluene	$\hat{E}$	$\hat{C}_2$	$\hat{\sigma}_v\ (xz)$	$\hat{\sigma}_v\ (yz)$
x	1	−1	1	−1
y	1	−1	−1	1
z	1	1	1	1
1	1	−1	1	−1
2	1	−1	−1	1
3	1	1	−1	−1
4	1	1	−1	−1
5	1	1	1	1
6	1	−1	1	−1
7	1	−1	−1	1
8	1	−1	1	−1
9	1	1	1	1
10	1	−1	1	−1
11	1	1	−1	−1
12	1	1	1	1
13	1	−1	1	−1
14	1	−1	−1	1
15	1	1	1	1
16	1	−1	−1	1
17	1	1	−1	−1
18	1	−1	1	−1
19	1	−1	−1	1
20	1	−1	1	−1
21	1	1	1	1
22	1	1	1	1
23	1	−1	−1	1
24	1	−1	−1	1
25	1	1	1	1
26	1	−1	−1	1
27	1	1	1	1
28	1	−1	−1	1
29	1	−1	1	−1
30	1	−1	−1	1
31	1	1	1	1
32	1	1	1	1
33	1	−1	−1	1
34	1	1	−1	−1
35	1	−1	−1	1
36	1	−1	−1	1
37	1	−1	−1	1
38	1	−1	−1	1
39	1	1	1	1

Table S5.

CCl_2F_2.

CCl_2F_2	$\hat{E}$	$\hat{C}_2$	$\hat{\sigma}_v\ (xz)$	$\hat{\sigma}_v\ (yz)$
x	1	-1	1	-1
y	1	-1	-1	1
z	1	1	1	1
1	1	1	1	1
2	1	1	-1	-1
3	1	-1	1	-1
4	1	1	1	1
5	1	-1	-1	1
6	1	1	1	1
7	1	-1	-1	1
8	1	1	1	1
9	1	-1	1	-1

Table S6.

C_2H_4.

C_2H_4	$\hat{E}$	$\hat{C}_2\ (z)$	$\hat{C}_2\ (y)$	$\hat{C}_2\ (x)$	$\hat{i}$	$\hat{\sigma}\ (xy)$	$\hat{\sigma}\ (xz)$	$\hat{\sigma}\ (yz)$
x	1	-1	-1	1	-1	1	1	-1
y	1	-1	1	-1	-1	1	-1	1
z	1	1	-1	-1	-1	-1	1	1
1	1	-1	1	-1	-1	1	-1	1
2	1	1	1	1	-1	-1	-1	-1
3	1	-1	-1	1	-1	1	1	-1
4	1	-1	1	-1	1	-1	1	-1
5	1	-1	-1	1	1	-1	-1	1
6	1	1	1	1	1	1	1	1
7	1	1	-1	-1	-1	-1	1	1
8	1	1	1	1	1	1	1	1
9	1	-1	-1	1	1	-1	-1	1
10	1	1	1	1	1	1	1	1
11	1	-1	1	-1	-1	1	-1	1
12	1	1	-1	-1	-1	-1	1	1

Table S7.

Toluene.

toluene	$\hat{E}$	$\hat{C}_2$	$\hat{\sigma}_v\ (xz)$	$\hat{\sigma}_{v'}\ (yz)$	transforms as	frequency
x	1	-1	1	-1		
y	1	-1	-1	1		
z	1	1	1	1		
1	1	-1	1	-1	x	273.5
2	1	-1	-1	1	y	398.2
3	1	1	-1	-1		400.7
4	1	1	-1	-1		497.1
5	1	1	1	1	z	569.4
6	1	-1	1	-1	x	570.1
7	1	-1	-1	1	y	709.1
8	1	-1	1	-1	x	826.4
9	1	1	1	1	z	833.2
10	1	-1	1	-1	x	900.7
11	1	1	-1	-1		1023.6
12	1	1	1	1	z	1071.7
13	1	-1	1	-1	x	1094.6
14	1	-1	-1	1	y	1098.5
15	1	1	1	1	z	1126.4
16	1	-1	-1	1	y	1146.4
17	1	1	-1	-1		1180.4
18	1	-1	1	-1	x	1210.9
19	1	-1	-1	1	y	1236.4
20	1	-1	1	-1	x	1239.7
21	1	1	1	1	z	1274.9
22	1	1	1	1	z	1321.3
23	1	-1	-1	1	y	1363.2
24	1	-1	-1	1	y	1507.5
25	1	1	1	1	z	1591.2
26	1	-1	-1	1	y	1608.5
27	1	1	1	1	z	1656.6
28	1	-1	-1	1	y	1669.2
29	1	-1	1	-1	x	1677.0
30	1	-1	-1	1	y	1715.6
31	1	1	1	1	z	1727.5
32	1	1	1	1	z	3015.9
33	1	-1	-1	1	y	3068.0
34	1	1	-1	-1		3069.0
35	1	-1	-1	1	y	3082.9
36	1	-1	-1	1	y	3225.4
37	1	-1	-1	1	y	3233.4
38	1	-1	-1	1	y	3251.8
39	1	1	1	1	z	3257.4

Table S8.
CCl_2F_2.

CCl_2F_2	$\hat{E}$	$\hat{C}_2$	$\hat{\sigma}_v\,(xz)$	$\hat{\sigma}_{v'}\,(yz)$	transforms as	frequency
x	1	−1	1	−1		
y	1	−1	−1	1		
z	1	1	1	1		
1	1	1	1	1	z	247.3
2	1	1	−1	−1		317.2
3	1	−1	1	−1	x	434.4
4	1	1	1	1	z	437.5
5	1	−1	−1	1	y	454.5
6	1	1	1	1	z	689.9
7	1	−1	−1	1	y	868.6
8	1	1	1	1	z	1388.1
9	1	−1	1	−1	x	1505.5

Table S9.
C_2H_4.

C_2H_4	$\hat{E}$	$\hat{C}_2\,(z)$	$\hat{C}_2\,(y)$	$\hat{C}_2\,(x)$	$\hat{\imath}$	$\hat{\sigma}\,(xy)$	$\hat{\sigma}\,(xz)$	$\hat{\sigma}\,(yz)$	transforms as	freq.
x	1	−1	−1	1	−1	1	1	−1		
y	1	−1	1	−1	−1	1	−1	1		
z	1	1	−1	−1	−1	−1	1	1		
1	1	−1	1	−1	−1	1	−1	1	y	823.6
2	1	1	1	1	−1	−1	−1	−1		844.6
3	1	−1	−1	1	−1	1	1	−1	x	1028.9
4	1	−1	1	−1	1	−1	1	−1		1077.4
5	1	−1	−1	1	1	−1	−1	1		1186.2
6	1	1	1	1	1	1	1	1		1370.9
7	1	1	−1	−1	−1	−1	1	1	z	1385.0
8	1	1	1	1	1	1	1	1		1786.5
9	1	−1	−1	1	1	−1	−1	1		3224.8
10	1	1	1	1	1	1	1	1		3251.4
11	1	−1	1	−1	−1	1	−1	1	y	3253.8
12	1	1	−1	−1	−1	−1	1	1	z	3259.7

Table S10.
Axis-transformation table for CH_3Br (matrices).

CH_3Br	$\hat{E}$	$\hat{C}_3$	$\hat{C}_3^2$	$\hat{\sigma}_v$	$\hat{\sigma}_{v'}$	$\hat{\sigma}_{v''}$
z	(1)	(1)	(1)	(1)	(1)	(1)
x,y	$\begin{bmatrix} 1 & 0 \\ 0 & 1 \end{bmatrix}$	$\begin{bmatrix} -\frac{1}{2} & -\frac{\sqrt{3}}{2} \\ \frac{\sqrt{3}}{2} & -\frac{1}{2} \end{bmatrix}$	$\begin{bmatrix} -\frac{1}{2} & \frac{\sqrt{3}}{2} \\ -\frac{\sqrt{3}}{2} & -\frac{1}{2} \end{bmatrix}$	$\begin{bmatrix} 1 & 0 \\ 0 & -1 \end{bmatrix}$	$\begin{bmatrix} -\frac{1}{2} & \frac{\sqrt{3}}{2} \\ \frac{\sqrt{3}}{2} & \frac{1}{2} \end{bmatrix}$	$\begin{bmatrix} -\frac{1}{2} & -\frac{\sqrt{3}}{2} \\ -\frac{\sqrt{3}}{2} & \frac{1}{2} \end{bmatrix}$

Table S11.
Axis-transformation table for CH_3Br (traces).

CH_3Br	$\hat{E}$	$\hat{C}_3$	$\hat{C}_3^2$	$\hat{\sigma}_v$	$\hat{\sigma}_{v'}$	$\hat{\sigma}_{v''}$
z	1	1	1	1	1	1
x,y	2	−1	−1	0	0	0

Requirement 9: Table S10 is the group representation table; **Table S11** is the group character table with group elements, rather than conjugacy class, listed across the top. The vibrations of CH_3Br labeled 641.7, 1286.2, and 3268.5 cm^{-1} transform in the same way as does the z-axis because the orientation of the molecule does not change under any of the symmetry operations. The three sets of doubly degenerate vibrations labeled 874.5, 1349.6, and 3180.3 cm^{-1} each include one vibration along the x-axis and another along the y-axis. Hence, you would expect to see all six peaks in the infrared spectrum of CH_3Br.

Requirement 10: Most chemistry departments have an infrared spectrometer, and most of these can handle gas phase samples. The chemicals that your chemistry department is most likely to have on hand are water, toluene, carbon dioxide, methane, and benzene; the other molecules (except allene) will be available in many schools. We recommend that the abstract algebra students actually take at least one spectrum of their own. The infrared spectra for the molecules listed in this ILAP are available at http://webbook.nist.gov/chemistry . To find these data, you need to search on a molecular name or formula; at the same time, you need to request gas phase IR spectrum.

Note that some spectra may consist of downward-going peaks (transmission of light spectra) while others may have upward-going peaks (absorption peaks). There is nothing wrong with either type of spectrum! The mathematical relationship between the intensities of the downward-going peaks ($T = I/I_0$) and those of the upward-going peaks (A) is $A = 10^{-T}$. It does not matter whether the peaks are downward- or upward-going; what matters is the frequencies at which the peaks appear. Note also that the intensities of the peaks in the spectra (how large I/I_0 is) are not considered in this ILAP. Some peaks are quite intense while others are quite weak.

Requirement 11: The possible chemicals are

- CH_3Br (pesticide from the strawberry field);

- CH_4 (from cow burps or farts, or from the natural gas pumping station);

- C_2H_4 (from the natural gas pumping station);

- CCl_2F_2 (from the air conditioner repair section of the auto repair shop), and

- toluene and/or benzene (from the gas station/auto repair shop).

The chemical responsible for the spectrum is CCl_2F_2. Therefore, this pollution should be attributed to the air conditioner section of the auto repair shop.

Students may answer this question by comparing the spectrum in **Figure 12** with those obtained in **Requirement 10**, or by comparison with their predicted spectra from **Requirements 8** and **9**. Encourage them to begin by comparing this spectrum with the ones they predicted in **Requirements 8** and **9**.

Note that you may change the mystery spectrum! Remember that the students have worked out synthetic spectra for methyl bromide (CH_3Br), water (H_2O), ethylene (C_2H_4), and toluene ($C_6H_5CH_3$).

Title: Pollution Police

Notes for the Instructor

In this project, students use linear algebra, group theory, and physical chemistry to detect atmospheric pollutants from infrared spectra. Students work with molecules found in such pollutants as pesticides, refrigerants, fuels, and products of fuel combustion. After using ball-and-stick models to determine the group of symmetry operations for each molecule, students compare the actions of the symmetry operations on the coordinate axes and on the vibrations of the molecule in order to identify the infrared light frequencies the molecule absorbs. Students use *Spartan* or a similar computer program found in most chemistry departments to view the molecular vibrations and to obtain the light absorption frequencies associated with these vibrations. Once they know which light frequencies should be absorbed by the various molecules, students determine if a spectrum taken with an infrared spectrometer corresponds to pollutants from a strawberry field (pesticide), cows, a natural gas pumping station, or an auto repair shop.

This ILAP is intended for use in a course in abstract algebra or linear algebra, or in the spectroscopy portion of an undergraduate or graduate physical chemistry course. Abstract algebra students need to know the definitions of group, abelian group, order of a group, and order of an element in a group, and should be familiar with groups of order up to 12. For linear algebra and physical chemistry classes, omit **Requirement 5**, which requires a knowledge of group theory. **Requirement 4**, concerning orders of symmetry operations, also may be omitted. Linear algebra students should have studied matrix representations of linear transformations of R^2 and R^3. Physical chemistry students who have not studied linear algebra may require additional explanation of some of the linear-algebraic terms in **Requirement 9**. Physical chemistry students' chemical training usually compensates for any lack of training in multivariable calculus and/or linear algebra.

The mathematics underlying modern spectroscopy is group representation and character theory, a field rarely covered in the undergraduate mathematics curriculum. Our goal in this project is to allow mathematics (and chemistry) students to see how linear algebra and group theory are used in chemistry—specifically, in spectroscopy—without having to build up the machinery of representation and character theory. Although using representation and character theory would give students a more accurate view of how abstract algebra is used in physical chemistry, we wanted the project to be accessible to any student in any first course in linear algebra or abstract algebra.

To use this project in mathematics courses, you will need to contact a physical chemistry colleague (preferably a spectroscopist) to share ball-and-stick molecular models, a molecular modeling program such as *Spartan*, and an infrared spectrometer. All chemistry departments have molecular models, most have modeling programs, and all have infrared spectrometers. We suggest that you team-teach this ILAP with your chemistry colleague so that she can help you with the ball-and-stick models (used in **Requirements 1, 3, 4,** and **6**), the modeling program (used in **Requirements 2, 7,** and **9**), and the infrared spectrometer (used in **Requirement 10**).

Because of the animated vibrational motions and the ability to rotate the molecules available in molecular modeling programs, we encourage you to make every effort to gain access to one of these programs. Infrared spectra are available at `http://webbook.nist.gov/chemistry`.

We recommend that you assign a variety of different molecules to students so that they examine both large and small molecules as well as molecules with different degrees of symmetry. However, our experience shows that student groups should not be assigned more than four molecules! Students cannot perform a complete analysis of sulfur trioxide, carbon dioxide, benzene, methane, allene, and sulfur hexafluoride, but they will work with water, ethylene, dichlorodifluoromethane, toluene, and methyl bromide throughout the project. Water is used as an example throughout the student problem statement.

Suggested division of molecules into three sets:

- Set 1: water, benzene, ethylene, sulfur trioxide

- Set 2: dichlorodifluoromethane, sulfur hexafluoride, methyl bromide

- Set 3: toluene, methane, allene

Suggested division of molecules into four sets:

- Set 1: water, methyl bromide, sulfur hexafluoride

- Set 2: dichlorodifluoromethane, allene, methane

- Set 3: toluene, sulfur trioxide

- Set 4: ethylene, benzene

Additional tips and suggestions for instructors are included in the **Sample Solution**.

Chemical Character Tables

Chemical character tables can be found in most physical chemistry textbooks; the most complete and widely used set can be found in Cotton [1990]. The following notation, known as the *Mulliken notation*, is used in chemical character tables.

- The major axis of rotational symmetry (the one with the largest n for C_n) is the z-axis.

- All rows whose values for the identity operation are 1 are labeled A or B depending on whether there is an indistinguishable transformation (indicated by +1) or a distinguishable transformation (indicated by -1), respectively, upon rotation about the C_n or S_n axes for that vibration. For example, the symmetric stretching and bending vibrations of water are both of A symmetry, while the asymmetric stretching vibration is of B symmetry.

- All rows whose traces for the identity operation $\hat{E}$ are 2 are labeled as E-type symmetries (doubly degenerate).

- All rows whose traces for the identity operation $\hat{E}$ are 3 are labeled as T-type symmetries (triply degenerate).

- A subscript of 1 indicates that the operations with respect to a C_2 axis perpendicular to the principal (highest order) axis C_n (or with respect to a σ) are symmetric, while a subscript of 2 indicates that these operations are antisymmetric.

- Subscripts u (*ungerade*, German for odd) and g (*gerade*, German for even) indicate parity upon inversion.

- For groups with a σ_h (plane of symmetry perpendicular to the principal axis), the superscript ' means that the row is symmetric with respect to reflection, while " means that it is not.

Further Reading

If you or your students wish to see how group representation and character theory are used in molecular spectroscopy, consult, for instance:

Bishop, David M. 1993. *Group Theory and Chemistry*. Mineola, NY: Dover.

Cotton, F. Albert. 1990. *Chemical Applications of Group Theory*. 3rd ed. New York: Wiley.

Hall, Lowell H. 1969. *Group Theory and Symmetry in Chemistry*. New York: McGraw-Hill

Herzberg, Gerhard. 1945. *Molecular Spectra and Molecular Structure: II. Infrared and Raman Spectra of Polyatomic Molecules*. Princeton, NJ: Van Nostrand.

McQuarrie, Donald A., and John D. Simon. 1997. *Physical Chemistry: A Molecular Approach*. Sausalito, CA: University Science Books.

Wilson, E. Bright, J.C. Decius, and Paul C. Cross. 1955. *Molecular Vibrations: The Theory of Infrared and Raman Vibrational Spectra*. New York: McGraw-Hill.

For a classroom spectroscopy project (ILAP) that makes more explicit use of group representation and character theoretic results, see

Halverson, Tom, and Tom Varberg. 1997. Molecular Vibrations and Symmetry. St. Paul, MN: Macalester College.

About the Authors

Jodye Selco, professor of chemistry at the University of Redlands since 1987, earned her B.S. in chemistry from the University of California, Irvine, in 1979 and her Ph.D. in chemistry from Rice University in 1984. She is a physical chemist specializing in spectroscopy (meaning that she uses a lot more mathematics than other chemists!). She enjoys performing her Chemistry Magic Show at local schools, star-gazing, growing all her own vegetables, and making long commutes on the Southern California freeways.

Janet Beery, professor of mathematics at the University of Redlands since 1989, earned her B.S. in mathematics and English literature form the University of Puget Sound in 1983 and her Ph.D. in mathematics from Dartmouth College in 1989, specializing in permutation group theory. She has a new-found interest in history of mathematics and currently is writing historical modules with high school teachers. Since her undergraduate days, she has enjoyed reading novels without having to write papers about them, as well as traveling to rainy locales.

Jodye Selco and Janet Beery both enjoy teaching with technology, writing and assigning classroom projects, and helping students discover ideas for themselves.

Guide for Authors

Focus

The UMAP Journal focuses on **mathematical modeling and applications of mathematics at the undergraduate level**. The editor also welcomes expository articles for the On Jargon column, reviews of books and other materials, and guest editorials on new ideas in mathematics education or on interaction between mathematics and application fields. Prospective authors are invited to consult the editor or an associate editor.

Understanding

A manuscript is submitted with the understanding—unless the authors advise otherwise—that the work is original with the authors, is contributed for sole publication in the *Journal*, and is not concurrently under consideration or scheduled for publication elsewhere with substantially the same form and content. Pursuant to U.S. copyright law, authors must sign a copyright release before editorial processing begins. Authors who include data, figures, photographs, examples, exercises, or long quotations from other sources must, before publication, secure appropriate permissions from the copyright holders and provide the editor with copies. The *Journal*'s copyright policy and copyright release form appear in Vol. 18 (1997) No. 1, pp. 1–14 and at `ftp://cs.beloit.edu/ math-cs/Faculty/Paul Campbell/Public/UMAP` .

Language

The language of publication is English (but the editor will help find translators for particularly meritorious manuscripts in other languages). The majority of readers are native speakers of English, but authors are asked to keep in mind that readers vary in their familiarity with vocabulary, idiomatic expressions, and slang. Authors should use consistently either British or American spelling.

Format

Even short articles should be sectioned with carefully chosen (unnumbered) titles. An article should begin by saying clearly what it is about and what it will presume of the reader's background. Relevant bibliography should appear in a section entitled *References* and may include annotations, as well as sources not cited. Authors are asked to include short biographical sketches and photos in a section entitled *About the Author(s)*.

Style Manual

On questions of style, please consult current *Journal* issues and *The Chicago Manual of Style*, 13th or 14th ed. (Chicago, IL: University of Chicago Press, 1982, 1993).

Citations

The *Journal* uses the author-date system. References cited in the text should include between square brackets the last names of the authors and the year of publication, with no intervening punctuation (e.g., [Kolmes and Mitchell 1990]). For three or more authors, use [Kolmes et al. 1990]. Papers by the same authors in the same year may be distinguished by a lowercase letter after the year (e.g., [Fjelstad 1990a]). A specific page, section, equation, or other division of the cited work may follow the date, preceded by a comma (e.g., [Kolmes and Mitchell 1990, 56]). Omit "p." and "pp." with page numbers. Multiple citations may appear in the same brackets, alphabetically, separated by semicolons (e.g., [Ng 1990; Standler 1990]). If the citation is part of the text, then the author's name does not appear in brackets (e.g., "...Campbell [1989] argued ...").

References

Book entries should follow the format (note placement of year and use of periods):

> Moore, David S., and George P. McCabe. 1989. *Introduction to the Practice of Statistics.* New York, NY: W.H. Freeman.

For articles, use the form (again, most delimiters are periods):

> Nievergelt, Yves. 1988. Graphic differentiation clarifies health care pricing. UMAP Modules in Undergraduate Mathematics and Its Applications: Module 678. *The UMAP Journal* 9 (1): 51–86. Reprinted in *UMAP Modules: Tools for Teaching 1988*, edited by Paul J. Campbell, 1–36. Arlington, MA: COMAP, 1989.

What to Submit

Number all pages, put figures on separate sheets (in two forms, with and without lettering), and number figures and tables in separate series. Send three paper copies of the entire manuscript, plus the copyright release form, and—by email attachment or on diskette—formatted and unformatted ("text" or ASCII) files of the text and a separate file of each figure. Please advise the computer platform and names and versions of programs used. The *Journal* is typeset in LaTeX using EPS or PICT files of figures.

Refereeing

All suitable manuscripts are refereed *double-blind*, usually by at least two referees.

Courtesy Copies

Reprints are not available. Authors of an article each receive two copies of the issue; the author of a review receives one copy; authors of a UMAP Module or an ILAP Module each receive two copies of the issue plus a copy of the *Tools for Teaching* volume. Authors may reproduce their work for their own purposes, including classroom teaching and internal distribution within their institutions, provided copies are not sold.

UMAP Modules and ILAPs

A UMAP Module is a teaching/learning module, with precise statements of the target audience, the mathematical prerequisites, and the time frame for completion, and with exercises and (often) a sample exam (with solutions). An ILAP (Interdisciplinary Lively Application Project) is a student group project, jointly authored by faculty from mathematics and a partner department. Some UMAP Modules and ILAPs appear in the *Journal*, others in the annual *Tools for Teaching* volume. Authors considering whether to develop a topic as an article, a UMAP Module, or an ILAP should consult the editor.

Where to Submit

Reviews, On Jargon columns, and ILAPs should go to the respective associate editors, whose addresses appear on the *Journal* masthead. Send all other manuscripts to

Paul J. Campbell, Editor
The UMAP Journal
Campus Box 194
Beloit College
700 College St.
Beloit, WI 53511–5595
USA
voice: (608) 363–2007   fax: (608) 363–2718   email: `campbell@beloit.edu`

Subscription Rates for 2001 Calendar Year: Volume 22

MEMBERSHIP PLUS FOR INDIVIDUAL SUBSCRIBERS

Individuals subscribe to *The UMAP Journal* through COMAP's Membership Plus. This subscription includes print copies of quarterly issues of *The UMAP Journal*, our annual collection *UMAP Modules: Tools for Teaching*, our organizational newsletter *Consortium*, on-line membership that allows members to search our on-line catalog, download COMAP print materials, and reproduce for use in their classes, and a 10% discount on all COMAP materials.

(Domestic)	#2120	$75
(Outside U.S.)	#2121	$85

INSTITUTIONAL PLUS MEMBERSHIP SUBSCRIBERS

Institutions can subscribe to the *Journal* through either Institutional Pus Membership, Regular Institutional Membership, or a Library Subscription. Institutional Plus Members receive two print copies of each of the quarterly issues of *The UMAP Journal*, our annual collection *UMAP Modules: Tools for Teaching*, our organizational newsletter *Consortium*, on-line membership that allows members to search our on-line catalog, download COMAP print materials, and reproduce for use in **any class taught in the institution, and** a 10% discount on all COMAP materials.

(Domestic)	#2170	$395
(Outside U.S.)	#2171	$415

INSTITUTIONAL MEMBERSHIP SUBSCRIBERS

Regular Institutional members receive only print copies of *The UMAP Journal*, our annual collection *UMAP Modules: Tools for Teaching*, our organizational newsletter *Consortium*, **and** a 10% discount on all COMAP materials.

(Domestic)	#2140	$165
(Outside U.S.)	#2141	$185

LIBRARY SUBSCRIPTIONS

The Library Subscription includes quarterly issues of *The UMAP Journal* and our annual collection *UMAP Modules: Tools for Teaching* and our organizational newsletter *Consortium*.

(Domestic)	#2130	$140
(Outside U.S.)	#2131	$160

To order, send a check or money order to COMAP, or call toll-free
1-800-77-COMAP (1-800-772-6627).

Second-class postage paid at Boston, MA
and at additional mailing offices.
Send address changes to:
The UMAP Journal
COMAP, Inc.
57 Bedford Street, Suite 210, Lexington, MA 02420